Studies in Economic Transition

General Editors: **Jens Hölscher**, Reader in Economics, University of Brighton; and **Horst Tomann**, Professor of Economics, Free University Berlin

This series has been established in response to a growing demand for a greater understanding of the transformation of economic systems. It brings together theoretical and empirical studies on economic transition and economic development. The post-communist transition from planned to market economies is one of the main areas of applied theory because in this field the most dramatic examples of change and economic dynamics can be found. The series aims to contribute to the understanding of specific major economic changes as well as to advance the theory of economic development. The implications of economic policy is a major point of focus.

Titles include:

Irwin Collier, Herwig Roggemann, Oliver Scholz and Horst Tomann (*editors*)
WELFARE STATES IN TRANSITION
East and West

Hella Engerer
PRIVATIZATION AND ITS LIMITS IN CENTRAL AND EASTERN EUROPE
Property Rights in Transition

Hubert Gabrisch and Rüdiger Pohl (*editors*)
EU ENLARGEMENT AND ITS MACROECONOMIC EFFECTS IN EASTERN EUROPE
Currencies, Prices, Investment and Competitiveness

Oleh Havrylyshyn
DIVERGENT PATHS IN POST-COMMUNIST TRANSFORMATION
Capitalism for All or Capitalism for the Few?

Jens Hölscher (*editor*)
FINANCIAL TURBULENCE AND CAPITAL MARKETS IN TRANSITION COUNTRIES

Jens Hölscher and Anja Hochberg (*editors*)
EAST GERMANY'S ECONOMIC DEVELOPMENT SINCE UNIFICATION
Domestic and Global Aspects

Mihaela Kelemen and Monika Kostera (*editors*)
CRITICAL MANAGEMENT RESEARCH IN EASTERN EUROPE
Managing the Transition

Emil J. Kirchner (*editor*)
DECENTRALIZATION AND TRANSITION IN THE VISEGRAD
Poland, Hungary, the Czech Republic and Slovakia

Tomasz Mickiewicz
ECONOMIC TRANSITION IN CENTRAL EUROPE AND THE COMMONWEALTH OF INDEPENDENT STATES

Julie Pellegrin
THE POLITICAL ECONOMY OF COMPETITIVENESS IN AN ENLARGED EUROPE

Stanislav Poloucek (*editor*)
REFORMING THE FINANCIAL SECTOR IN CENTRAL EUROPEAN COUNTRIES

Gregg Robins
BANKING IN TRANSITION
East Germany after Unification

Johannes Stephan
ECONOMIC TRANSITION IN HUNGARY AND EAST GERMANY
Gradualism and Shock Therapy in Catch-up Development

Johannes Stephan
TECHNOLOGY TRANSFER VIA FOREIGN DIRECT INVESTMENT IN CENTRAL AND EASTERN EUROPE
Theory, Method of Research and Empirical Evidence

Hans van Zon
THE POLITICAL ECONOMY OF INDEPENDENT UKRAINE

Adalbert Winkler (*editor*)
FINANCIAL DEVELOPMENT IN EASTERN EUROPE
The First Ten Years

Studies in Economic Transition
Series Standing Order ISBN 0–333–73353–3
(outside North America only)

You can receive future titles in this series as they are published by placing a standing order. Please contact your bookseller or, in case of difficulty, write to us at the address below with your name and address, the title of the series and the ISBN quoted above.

Customer Services Department, Macmillan Distribution Ltd, Houndmills, Basingstoke, Hampshire RG21 6XS, England

Technology Transfer via Foreign Direct Investment in Central and Eastern Europe

Theory, Method of Research and Empirical Evidence

Edited by
Johannes Stephan

palgrave
macmillan

First published 2006 by
PALGRAVE MACMILLAN
Houndmills, Basingstoke, Hampshire RG21 6XS and
175 Fifth Avenue, New York, N.Y. 10010
Companies and representatives throughout the world

PALGRAVE MACMILLAN is the global academic imprint of the Palgrave Macmillan division of St. Martin's Press, LLC and of Palgrave Macmillan Ltd. Macmillan® is a registered trademark in the United States, United Kingdom and other countries. Palgrave is a registered trademark in the European Union and other countries.

ISBN-13: 978–1–4039–9952–8
ISBN-10: 1–4039–9952–X

This book is printed on paper suitable for recycling and made from fully managed and sustained forest sources.

A catalogue record for this book is available from the British Library.

Library of Congress Cataloging-in-Publication Data
Technology transfer via foreign direct investment in Central and Eastern Europe : theory, method of research and empirical evidence / edited by Johannes Stephan.
p. cm.
Includes bibliographical references.
ISBN 1–4039–9952–X (cloth)
1. Investments, Foreign–Europe, Eastern. 2. Investments, Foreign–Europe, Central. 3. Technology transfer–Europe, Eastern. 4. Technology transfer–Europe, Central. I. Stephan, Johannes, 1966–
HG5430.7.A3T43 2006
338.947'026–dc22 2006051475

10 9 8 7 6 5 4 3 2 1
15 14 13 12 11 10 09 08 07 06

Printed and bound in Great Britain by
Antony Rowe Ltd, Chippenham and Eastbourne

Contents

Part II Empirical Studies

List of Tables

List of Figures

Preface

The analysis of this book is guided by a interest in the question of to what extent foreign direct investors transfer technology or knowledge to their affiliates, and whether there are any external effects for domestic firms in the transition economies. This interest is embedded in the general research programme on Conditions of Economic Development in Central East Europe which gave rise to the Palgrave's Studies in Economic Transition series, edited by Jens Hölscher and Horst Tomann.

Foreign direct investment (FDI) plays a particularly important role for technological development if investment flows from higher-developed economies to countries or regions with a lower level of technology. According to Dunning's 'OLI paradigm' (see, e.g., Dunning, 1981), FDI takes place only if three conditions are met: first, the firm must possess some form of *ownership advantages* (including technology); second, the host economy must offer some *locational advantages*; third, the investor has to be able to *internalise* the benefits to the host economy derived from his proprietary knowledge (for the latter condition, see also Markusen, 1984). This suggests that FDI is potentially associated with technology transfer to the host economy. Technology and knowledge can be transferred in a variety of ways, such as product and process technology (either embedded in products or machinery, or in the form of blue-prints), management practices and expertise (see, e.g., Findlay, 1978, for a model on the latter two, or Dyker, 1999 for a conceptualization including different theoretical approaches), intensified competition (see, e.g., Blomström and Kokko, 1996; Markusen and Venables, 1997) and finally as information about and access to foreign markets (see, e.g., Rasiah, 1995).

The choice of transition countries in Central and Eastern Europe (CEE) as the subjects of technology transfer research in this volume offers some particularly insightful experiences: post-communist countries share a number of transition-specific conditions, and are therefore comparable as a group and to a certain extent can be differentiated from developing and developed countries. Campos and Kinoshita (2002) argue that transition economies are far away from the international technological frontier and hence can potentially benefit from technology transfer to a significant extent. Yet, unlike many developing countries, they started out with a long history of industrialisation and a relatively well-educated workforce. Moreover, these economies are in proximity to the richer and more developed Western European markets and most transition economies embarked on a comprehensive privatisation process at the time when FDI was starting to peak on a world-wide scale (Campos and Kinoshita, 2002: 3). Some authors even state that the transfer of technology and know-how accompanying FDI is more important for transition economies than capital transfers (McMillan, 1996; Hunya, 1998).

FDI has played a significant role in the enterprise restructuring of the transition countries in CEE. CEE governments also provide incentives for FDI. Justifications for these special incentives have traditionally been the possible beneficial effects caused by the transfer of technology from the parent company to its local affiliate and the related positive spillover effects to the domestically-owned firms of the host country. We can safely assume that the economic activity of a foreign investor will typically help to accelerate technological development in the host economy to some degree (for literature reviews see, e.g., Lim, 2001; Dyker and Stolberg, 2003; for CEE see, Hunya, 2000). However, our knowledge about the actual *channels* and *intensity* of knowledge transfer – and in particular the conditions for intense transfer – is still scarce. The large body of econometric evidence produces an inconclusive picture which is partly related to the selection of methods of analysis (see Görg and Greenaway, 2002). On the other hand, single case studies are typically limited by the lack of generality of their results.

The research on the link and the causality between FDI inflows and economic growth at the aggregate level for the CEE region is inhibited by transition-specific particularities in the growth accounting approach (see on this also Havrylyshyn, 2001). This book therefore employs a microeconomic as well as a comparative industry perspective on technology transfer via foreign subsidiaries in transition countries. This volume aims to contribute to the discussion by using state-of-the art econometric research and by introducing an innovative 'mezzo-case' study approach. The latter is interdisciplinary in as much as it draws from international business management research and involves a large number of foreign investment enterprises (FIEs) in CEE. Part II, the empirical part of this volume is preceded in part I by a comprehensive yet concise overview of what scholars have up to now established in terms of technology transfer via FDI, with particular reference to CEE.

This review, edited by Björn Jindra, includes a synthesis of theory, an appraisal of methodology and an analysis of current empirical evidence. Jindra consolidates the body of theoretical literature on technology transfer by classifying the channels and determinants of internal and external effects of knowledge transfer via FDI. He also discusses a variety of methodological problems involved in technology transfer research. A meta-analysis is conducted to amalgamate the empirical findings from the relevant econometric studies. This is supplemented by scrutinising the first findings from the emerging mezzo-analysis. The body of empirical evidence is then compared to the general theory in order to map out future research trends. It emerges that there is clear evidence for direct technology transfer from the parent companies to the foreign subsidiaries in CEE. However, the hard evidence for spillover effects to the respective domestic economy is sobering (Rodrik 1999), despite some indications for the importance of backward linkages to domestic suppliers. All types of technology transfer via FDI are subject to a variety of host country-, industry-, multinational corporation (MNC)- and subsidiary-specific conditions. Part I of the book comprises Chapters 1–5.

Part II of the book then presents the results of two particularly interesting pieces of analysis, one based on the methodology of econometric testing and the other on large-scale case studies in a mezzo-approach. In an introductory chapter to part II, by Judit Hamar and Johannes Stephan, the most important characteristics of FDI into CEE are assessed in a comparative methodology for the countries of Estonia, Poland, the Slovak Republic and Slovenia, with a particular emphasis on the special case of Hungary.

As an example of an up-to-date econometrics analysis on the subject-matter at the research–technological frontier area, Priit Vahter then tests technology transfer in two particularly significant small countries in the region – Estonia and Slovenia. The emphasis is on the export or local market orientation of FIEs as a determinant of technology transfer. The study is based on firm-level panel data. It is shown that, in Estonia, export oriented FIEs have on average much lower labour productivity levels than domestic market oriented foreign affiliates. In Slovenia, on the contrary, the export orientation of a foreign affiliate is not correlated with lower labour productivity. No horizontal spillovers of FDI to domestic firms can be detected in Estonia. In Slovenia, however, positive spillovers to domestic firms and other affiliates have been identified. The findings show that different types of FDI can have different effects on the host country and that the existence of positive spillovers may depend on the level of economic development of that country. This analysis set out in Chapter 6. As an example of an innovative approach of mezzo-analysis, Judit Hamar and Johannes Stephan report in Chapter 7 on their analysis of a unique database on foreign investor subsidiaries' strategies and mandates, generated in questionnaire-based fieldwork during 2002 and 2003. Using an interdisciplinary approach including international business and strategic management literature, the analysis compares the potentials for intense and dynamic technology transfer across a selection of countries including Estonia, Hungary, Poland, the Slovak Republic and Slovenia. Among the most important findings, they conclude that while the FIEs in Hungary are particularly adept at benefiting from the large potential of technology and knowledge transfer, they exhibit only little contact with the host economy – i.e. domestic firms are unlikely to benefit from technology transfer to any great extent. In all other countries, adaptive abilities were rather less developed, either because the subsidiaries were already and rather prematurely autonomous in their decisions on business functions (as appears to be the case for the rather market oriented Polish FIEs), or are still too young and immature and are more closely controlled by their foreign investors (as seems to be the case for the Slovak subsidiaries). The case study analysis closes with a regression model on the determinants of FIE development, which formed part of an EU-funded research project.

This design of this book, comprising two parts and several major contributions, offers a particular insight into the subject matter: after an appraisal of theory and existing empirical research, two examples of empirical analysis using a completely different methodology should permit the reader to see the most up-to-date knowledge that science has to offer about the role of FDI in CEE.

Acknowledgements

The analysis presented in this book has been carried out by Björn Jindra, Judit Hamar, Johannes Stephan and Priit Vahter. Björn Jindra is affiliated to University College, London University (OCL) and the Halle Institute for Economic Research, Germany, and was mainly responsible for Part I of the book (Chapter 1–5). Judit Hamar is a senior researcher at the Kopint-Datorg Foundation for Economic Research, Budapest, Hungary and Johannes Stephan is senior researcher at the Halle Institute for Economic Research, Germany. The latter two were mainly responsible for the fieldwork study (Chapter 7) and the introductory chapter for Part II of this book (Chapter 6). Priit Vahter is affiliated to the University of Tartu, Estonia, and the Bank of Estonia and was mainly responsible for the econometric study on Estonia and Slovenia (Chapter 7).

The fieldwork study formed part of a larger research project supported by the EU Commission in the form of an RTD project in the 5th Framework Programme in the Key Action on Improving the Socio-Economic Knowledge Base, in the workpackage entitled 'Mapping the Technology Structure of Branch Plants and Technology Integration of CEECs'. The project 'EU Integration and the Prospects for Catch-Up Development in CEECs – The Determinants of the Productivity Gap' (contract no HPSE-CT-2001-00065, www.ihw-halle.de/projects/productivity-gap.htm) was concerned with the determinants of productivity gaps between Eastern and Western Europe and was conducted between September 2001 and August 2004 (Disclaimer: the authors are solely responsible for the contents, which may not represent the opinion of the Community. The Community is not responsible for any use that may be made of data appearing in this publication.)

Thanks go to the numerous researchers refereeing earlier versions of the analysis presented here who often served as external experts to the larger EU research project. Here, in particular, Michael Kaser (St Antony's College, Oxford) provided scientific guidance. Particular thanks go to Igor Filatotchev (King's College University of London) and Tomasz Mickiewicz (UCL), the intellectual fathers of the methodology for the mezzo field study; Slavo Radošević (UCL), who designed the conceptual framework for this fieldwork and who supervised Björn Jindra's analysis in the framework of an MA thesis; Urmas Varblane

(University of Tartu, Estonia) as supervisor of the analysis of Priit Vahter's work in the framework of his MA thesis; Attila Béres for assistance in the calculations in Chapter 6 of the book and the fieldwork study (Chapter 7), and Ádám Mészáros (PhD student) who contributed to the Hungarian country study from which Chapter 6 was distilled. Both are affiliated to Kopint-Datorg, Hungary. The contribution of the other project participants involved in the fieldwork – namely Helena Hannula (University of Tartu, Estonia), Boris Majcen (Institute for Economic Research, Ljubljana), Katrin Männik (University of Tartu, Estonia), Romuald Niedzielski (University of Opole, Poland), Matija Rojec (Institute for Economic Research, Ljubljana), Tomas Sabol (University of Kosice, Poland), Vincent Šoltés (University of Kosice, Poland) and Michal Šoltés (University of Kosice, Poland), have all had important inputs into this book.

Notes on the Contributors

Judit Hamar is a graduate of Budapest University of Economics, where she was awarded her first doctorate in International Economics and her second in Corporate Management. She has a very long and outstanding employment record including international placements at the University of Lille, Université de Paris (I et X), University of Pennsylvania, Universita' Degli Studi di Verona, The Queen's University of Belfast. Today, she is Chief Researcher at Kopint-Datorg in Budapest, and her key expertise expand from international economics, foreign trade and foreign investment, competitiveness, corporate structure and performance, international manufacturing networks, to the issue of small and medium enterprises. Judit Hamar has participated and led various research projects on the issue of technology transfer and FDI, including from National bodies and the EU (PHARE-ACE), UN, and the World Bank.

Bjöern Jindra is a graduate of the University of Wolverhampton, the University of Brno, and University College London, where he was awarded his Master of Arts in Political Economies of Russia and Eastern Europe in 2004. Today, he is researcher and PhD student at the Halle Institute for Economic Research, where he focuses his research on real convergence, technological development and innovation, and technology transfer with a particular focus on Central and East Europe.

Johannes Stephan is a graduate of the Free University Berlin and was awarded his PhD at the Institute for German Studies, University of Birmingham, in 1997. Upon completion of this project, he was appointed full time researcher to the Halle Institute for Economic Research to focus on EU-integration and real economy development in Central and East Europe. There, he was able to secure funding for a larger three year, international collaborative research project from the EU Commission in its 5th Framework Programme as project co-ordinator, and for a one-year project in the 6th Framework Programme as workpackage leader. Today, he is head of the Department for Industrial Organisation and Regulation Economics at the Halle Institute for Economic Research, and his research interests are focussed upon the economics of integration, systemic transformation, and economic catch-up development.

Priit Vahter is a graduate of the University of Tartu, where he was awarded his Master of Arts in Economics in 2004. Today, he is researcher and PhD student at the Central Bank of Estonia, Tallinn, where he devotes his particular interest to technology transfer studies of host and home country effects of foreign direct investments and the determinants of productivity growth, including innovation.

List of Abbreviations

CEE	Central and Eastern Europe
CEEC	Central and Eastern European country
DE	Domestic enterprise
EU-15	EU members before the 2004 enlargement
EUROSTAT	European Statistical Body
FDI	Foreign direct investment
FE	Fixed effects
FIE	Foreign investment subsidiary
FISIM	Financial Intermediation Services Indirectly Measured
FPE	Former planned economy
GDP	Gross domestic product
GFCF	Gross fixed capital formation
GMM	General method of moments
HI	Herfindahl index
HSO	Hungarian Statistical Office
IMF	International Monetary Fund
IWH	Walle Institute for Economic Research
IP	Intellectual property
ISO	International Statistical Office
JV	Joint venture
LS	Least squares
M&A	Mergers and acquisitions
MNC	Multinational corporation
MNE	Multinational enterprise
NACE	Classification of Economic Activities in the European Union
OECD	Organisation for Economic Cooperation and Development
OLS	Ordinary least squares
OPT	Outward processing trade
PF	Production frontier
PPI	Producer price index
PPP	Purchasing power parity
R&D	Research and development
RE	Random effects

SME	Small- and medium-sized enterprise
SOE	State-owned enterprise
SR	Solow residual
TFP	Total factor productivity
TNC	Transnational corporation
WIIW	Vienna Institute for International Economic Studies

Introduction

The analysis of this book is guided by the interest in the question to which extent foreign direct investors transfer technology or knowledge to their affiliates, and whether there are any external effects to domestic firms in the transition economies. This interest is embedded in the general research programme of Conditions of Economic Development in Central East Europe which gave rise to the series 'Studies in Economic Transition' edited by Jens Hölscher and Horst Tomann.

Foreign direct investment plays a particularly important role for technological development if investment flows from higher developed economies to countries or regions with a lower level of technology. According to Dunning's 'OLI paradigm' (see, e.g., Dunning, 1981), foreign direct investment takes place only if three conditions are met: first, the firm must possess some form of ownership advantages (including technology); second, the host economy must offer some locational advantages; third, the investor has to be able to internalise the benefits to the host economy derived from his proprietary knowledge (for the latter condition, see also Markusen 1984). The first condition suggests that FDI is potentially associated with technology transfer to the host economy. (not, however, the third condition, because it would not allow any external effects to materialise, and internal effects can potentially be appropriated abroad; hence no effect at all for the host economy). Technology and knowledge can be transferred to the host economy via foreign direct investment in the form of product and process technology (either embedded in products or machinery, or in the form of blue-prints), in the form of management practices and expertise (see, e.g., Findlay, 1978, for a model on the latter two, or Dyker, 1999 for a conceptualisation including different theoretical approaches), in the form of intensified competition (see, e.g., Blomström/Kokko, 1996, and Markusen/Venables, 1997), and finally as information about and access to foreign markets provided by the foreign parent (see, e.g., Rasiah, 1995).

The choice of transition countries in Central East Europe (CEE) as subject to technology transfer research in this volume offers some particularly insightful experiences: post-communist countries share a number of transition specific conditions, and are therefore comparable as a group and to a certain extent can be differentiated from developing countries and developed countries. Campos and Kinoshita (2002) argue that transition economies were far away from the international technological frontier and hence could potentially benefit from technology transfer to a particularly large extent. Yet, in distinction to many developing countries, they started out with a long history of industrialisation and a relatively well educated workforce. Moreover, these economies are in proximity to the richer and more developed Western European markets and most transition economies embarked on a comprehensive privatisation process at the time when foreign direct investment (FDI) was starting to peak on a world-wide scale (p. 3). Some authors even state that the transfer of technology and know-how accompanying FDI is already more important for transition economies than the sole capital transfer (Hunya, 1998, McMillan, 1996).

FDI has had a significant role in enterprise restructuring of transition countries in CEE. Governments in CEE also provide a lot of incentives for FDI. Justifications for these special incentives are traditionally the possible beneficial effects caused by the transfer of technology from the parent company to its local affiliate and the related positive spillover effects to the domestic owned firms of the host country. We can safely assume that economic activity of a foreign investor will typically help to accelerate technological development in the host economy to some degree (for literature reviews, see e.g. Dyker/Stolberg, 2003, or Lim, 2001, and for CEE: Hunya, 2000). However, our knowledge about the actual channels and intensity of knowledge transfer – and in particular the conditions for intense transfer – is still scarce. The large body of econometric evidence produces an inconclusive picture, which partly roots in the selection of methods of analysis (see Gorg/Greenaway, 2002). On the other hand, single case studies are typically limited by the lack of generality of results.

This book aims to contribute to the discussion of technology transfer via FDI in CEE by using state-of-the-art econometric research and by introducing an innovative mezzo case study approach. The latter is interdisciplinary in as much as it draws from international business management research and involves a large number of foreign investment subsidiaries (FIEs) in CEE. The empirical part of this volume is preceded by a comprehensive yet concise overview of what scholars have until now established in terms of technology transfer via foreign direct investment with particular reference to Central East Europe. This review, edited by Björn Jindra (University College London), includes a synthesis of theory, appraisal of methodology, and analysis of current empirical evidence. The author consolidates the body of theoretical literature on technology transfer by classifying channels and determinants of internal and external effects of knowledge transfer via FDI. He discusses a variety of methodological problems involved in technology transfer research. A meta-analysis is conducted to amalgamate the empirical findings from relevant econometric studies. This is supplemented by scrutinising the first findings from the emerging mezzo-analysis. The body of empirical evidence is compared to the general theory in order to map out future research trends. The first part of the book comprises chapters one to four and is deducted from his excellent MA thesis.

The second part of the book then presents the results of two particularly interesting pieces of analysis, one based on the methodology of econometric testing, and the other based on large-scale case studies in a mezzo-approach. In an introductory chapter to part two of this book, the most important characteristics of FDI into CEE are assessed in a comparative methodology for the countries of Estonia, Poland, the Slovak Republic, Slovenia, and with a particular emphasis on the special case of Hungary.

As an example of an up-to-date econometrics analysis on the subject-matter at the research-technological frontier area, Priit Vahter (University Tartu, Estonia) tests technology transfer in two particularly insightful small countries in the region, namely Estonia and Slovenia. The emphasis is on export or local market orientation of foreign investment enterprises as a determinant of technology transfer. The study is based on firm-level panel data. It is shown that in Estonia, export oriented foreign investment enterprises have, on average, much lower labour productivity levels than domestic market oriented foreign affiliates. In Slovenia, on the contrary, the export orientation of a foreign affiliate is not correlated with lower labour productivity. No horizontal spillovers of foreign direct investment to domestic firms are detected in Estonia. In Slovenia, however, positive spillovers to domestic firms and other affiliates have been identified. The findings show that different types of foreign direct investment can have different effects on the host country, and that the existence of positive spillovers may depend on the level of economic development of the host country. This analysis is deduced from his award-winning MA thesis reproduced in chapter 6. As an example of an innovative approach of mezzo-analysis, Judit Hamar (Kopint-Datorg) and Johannes Stephan (Halle Institute for Economic Research) report on their analysis of a unique database on foreign investor subsidiaries' strategies and mandates, generated in questionnaire-based field work during 2002 and 2003. Using an interdisciplinary approach including international business and strategic management literature, the analysis compares the potentials for intense and dynamic technology transfer across a selection of countries including Estonia, Poland, the Slovak Republic, Hungary and Slovenia. Amongst the most important findings, they conclude that whilst the foreign investment subsidiaries (FIEs) in Hungary are particularly adept to benefit from large potentials of technology and knowledge transfer, they exhibit only little contact with the host economy, i.e. domestic firms are unlikely to benefit from technology transfer to any great extent. In all other countries, adaptive abilities were rather less developed, either because the subsidiaries are already and rather prematurely autonomous in decisions over business functions as appears to be the case for the rather market-oriented Polish FIEs, or are still too young and immature and are more closely controlled by their foreign investors as is suggested for the Slovak subsidiaries. The case-study analysis closes with a regression model on the determinants of FIE development. This analysis formed part of an EU-funded research project and is presented in chapter seven of this book.

The design of this book, comprising of two parts and three major contributions, offers a particular insight into the subject-matter: after an appraisal of theory as well as of existing empirical research, two examples of empirical analysis of completely different methodology should allow the reader to gain the most up-to-date knowledge that science has to offer about the role of FDI in CEE.

Part I

Theory and Review of the Latest Research on the Effects of FDI into Central East Europe

Part I

1
Introduction: The Scope of the Review

Björn Jindra

This review of the existing literature on the effects of FDI on the economies of CEE employs a microeconomic perspective on technology transfer via foreign subsidiaries in transition countries. It thus does not include the nexus between FDI, technology transfer and economic growth – i.e. the link between FDI inflows and economic growth at the aggregate level. The reason for this decision is based on the fact that economic growth analysis of post-communist economies is extremely difficult. First, only a short period of time is available for analysis, which is unfavourable in growth accounting econometric studies. Second, current research in this area shows an overwhelming consensus that traditional factor inputs, including investment, have no role in explaining growth (Havrylyshyn, 2001). Apart from a study by Campos and Kinoshita (2002), FDI is found in most studies not to be significant as an economic growth-enhancing factor. This is because in the transition-specific context reform policies seem to be the most important explanatory variables (Havrylyshyn, 2001).

This review contrasts a detailed theoretical framework of technology transfer via FDI with firm-level evidence from empirical studies in transition countries. As there is no comprehensive theory of technology transfer, the first task is to formulate a synthesis of general theoretical thought on technology transfer in respect to terminology, the general channels of international technology transfer, the explicit channels of technology transfer via FDI and their possible impacts and determinants. In our analysis, we distinguish between *direct*, *vertical* (forward, backward), *horizontal* and *regional* types of technology transfer via inward FDI. We propose a classification of determinants into host country-, industry-, MNC- and foreign subsidiary-specific factors related to each type of technology transfer.

The analysis scrutinises the empirical evidence with the help of research questions such as:

- Does technology transfer take place?
- What is the impact and what are the dynamics of technology transfer?
- What are the significant technology channels and the significant determinants for each type of technology transfer?

The bulk of empirical work has been produced by econometric studies employing the production function approach. After discussing the methodological challenges involved, a simple form of meta-analysis is conducted with a sample containing nineteen studies across twelve transition countries during the 1990s. The meta-analysis shows that direct effects are the most important channel for technology transfer via FDI in transition countries. Horizontal effects are found not to be significant in most cases. Only a few studies test for vertical effects and find positive backward linkages, which are larger than horizontal effects where they exist. Forward linkages tend to be negative. Regional effects exist independent from external effects at the national level. This pattern of technology transfer mirrors the findings from other regions. It is fair to say that the more advanced transition countries such as the Czech Republic, Hungary, Poland and Slovenia tend to show strong evidence for positive direct effects, which is likely to be coupled with positive backward linkages. This seems not to be the case for the less advanced countries such as Bulgaria and Romania. Traditional econometric studies do not offer insights into the explicit channels of technology transfer via FDI. The coverage of determinants for technology transfer tested is far less than the list of factors developed by the theoretical literature. In particular MNC- and subsidiary-specific factors are insufficiently explored.

This chapter also looks at the newly emerging research from mezzo-analysis, which includes *qualitative data* in the analysis. The respective studies find that MNCs operating in Central and Eastem European Countries (CEECs) seem to prefer a headquarter-centred approach to foster the growth of their affiliates. A number of subsidiary-specific determinants are found to be relevant for direct technology transfer. Yet a variety of factors in respect to direct as well as vertical technology transfer need further consideration.

Rodrik (1999) argues that 'today's policy literature is filled with extravagant claims about positive spillovers from FDI, but the hard evidence is sobering' (1999: 37). The findings of our analysis support this claim. Much theoretical and empirical work has been dedicated to spillover effects. However, direct technology transfer dominates. This raises serious questions:

- Is there productivity convergence or divergence between MNCs and domestic firms?
- To what extent and how do consumers and domestic producers benefit from FDI, if most of the effects are appropriated by the MNC or its foreign affiliate?

The chapter is organised as following way. The first section introduces the theoretical framework and is subdivided into parts dealing with terminology, international technology transfer, technology transfer via inward FDI and the determinants of technology transfer types. Also, three basic research questions are outlined. The next section considers empirical evidence from the transition countries. In the first part methodological challenges are explored, following a meta-analysis of relevant studies. The second part considers the empirical evidence from the field of mezzo-analysis. The final section concludes, suggests possible paths for the enhancement of current research and indicates some emerging research trends.

2
The Theoretical Framework: FDI and Technology Transfer

Björn Jindra

Chapter 2 introduces the theoretical framework which has developed around the topic of technology transfer via FDI. After some basic definitions and a short overview of the theory of the multinational enterprise (MNE), this chapter focuses on the channels and determinants of technology transfer via inward FDI. At the end of the chapter a classification of determinants into host country-, industry-, MNC- and subsidiary-specific factors in relation to technology transfer types is developed.

FDI, foreign subsidiary and MNC

Unlike portfolio- investment, FDI is a *long-term oriented investment abroad* with the main objective of the investor being to gain a significant impact on the company's *decision making processes* (Krugman and Obstfeld, 2000: 169). According to the International Monetary Fund (IMF), a significant impact is possible when the foreign investor holds a share of at least 10 per cent of the nominal capital. Lower shares count as *portfolio investment*, with the main objective of the investor being to realise short-term gains (IMF, 1993). It follows that FDI can be distinguished from portfolio investment due to its stronger commitment of investors to the host economy.

FDI materialises in the form of *foreign subsidiaries*. The OECD (1999) defines a foreign subsidiary as a company of which more than 50% of the voting shares are owned by another corporation abroad, termed the 'parent company' (OECD, 1999: 72). In the host economy foreign subsidiaries become real through the acquisition of an existing company or the foundation of a new firm (greenfield investment). Companies that establish operating units via FDI in at least two countries are labelled 'multinational companies' (MNCs). MNCs also conduct minority foreign investments and joint ventures (JVs), where the effective management of the foreign investment enterprise is shared with one or more partners, who can be local or foreign. In this chapter the term 'foreign subsidiary' is used to embrace all companies having at least 10% foreign capital, which is the most common threshold used in empirical studies.[1]

MNEs

Why do firms internationalise, and under what conditions do MNCs choose to set up production facilities overseas rather than export directly and/or licence their product: These questions have been a concern in the international business literature for a long time. There were different lines of explanation[2] which, however, did not initially constitute a theory of the MNC. This arguably started only with the internalisation literature drawing on authors such as Hymer and Coase. Hymer (1976) argues that MNCs invest abroad to reduce competition and to increase barriers to entry by establishing *collusive networks*. As internationalising MNCs face certain higher cost abroad compared to local domestic enterprises they must have some form of firm-specific advantage such as superior technology or cost advantage due to economies of scale. Another line of explanation goes back to Coase (1937), who argues that a firm exists when the transaction costs of markets exceed the cost of internalisation or establishing a firm hierarchy. It could therefore be advantageous for the MNC to internalise externalities resulting from market imperfections such as risk and uncertainty related to market transaction. Dunning (1977, 1981) argues that the internalisation approach alone cannot explain the level, structure and location of *all* international production. He organised the conditions for production abroad in his so called 'OLI paradigm'. It emphasises three circumstances:

- First, the firm must possess some *ownership advantages* such as a product, a production process, technology, reputation or other intangible assets, which allow the firm to exploit several markets.
- Second, the host economy must offer some *locational advantages* such as customer access, lower taxes and wages, or tariff avoidance, implying that production in more than one country is efficient.
- However, as it could still be more efficient to have a local company within the foreign market produce via licence-agreement or to

export, a third condition must be met before a foreign subsidiary will be established. There should be *internalisation considerations* conferring an advantage in having the production done within a single firm rather than by many firms or at arm's length through markets.

Technology transfer

Technology is an abstract and broad concept. The literature uses the terms 'knowledge' and 'technology' interchangeably. Blalock and Gertler (2004) define 'technology' to mean the managerial practices, production methods and other tacit and codified know-how by which a firm transforms capital, labour and materials into a product. From the notion of the firm-specific advantage discussed above, it seems that MNCs do not only supply capital that mobilises labour and land productively: they can also act as *conduits of technology transfer*. It is therefore generally assumed that internationalising firms operate at a higher technological level compared to companies in the host economy. Dunning's OLI paradigm does not explain how technology is transferred across borders. However, the process of internationalisation implies that technology as a form of ownership advantage can to a certain extent be transferred to the production facility abroad – i.e. to the foreign subsidiary.

General channels for technology transfer

Before we analyse the specific channels for technology transfer via inward FDI, we have to be aware of other possible channels for international technology transfer (see Figure 2.1, left-hand side). A first channel is direct transfer via *international licensing agreements* (Eaton and Kortum 1996), though recently it has been argued that these provide a less important source as the latest and most valuable technologies are not available on license (*World Investment Report*, 2000).

The second channel is *international trade*, in particular through the importing of intermediate products and capital equipment (Markusen, 1989; Grossman and Helpman, 1991; Feenstra, Markusen and Zeile 1992) as well as through learning by exporting into industrial countries (Clerides, Lach and Tybout 1997). It is important to note that large portions of global trade are between MNCs and between MNCs and their foreign subsidiaries (Blomström and Kokko, 1996; Meyer, 2003). The third channel embraces *non-equity inter-firm relationships* such as franchising, subcontracting, strategic alliances, which again involve MNCs to a large extent (Meyer, 2003: 12). Van Pottelsberghe and

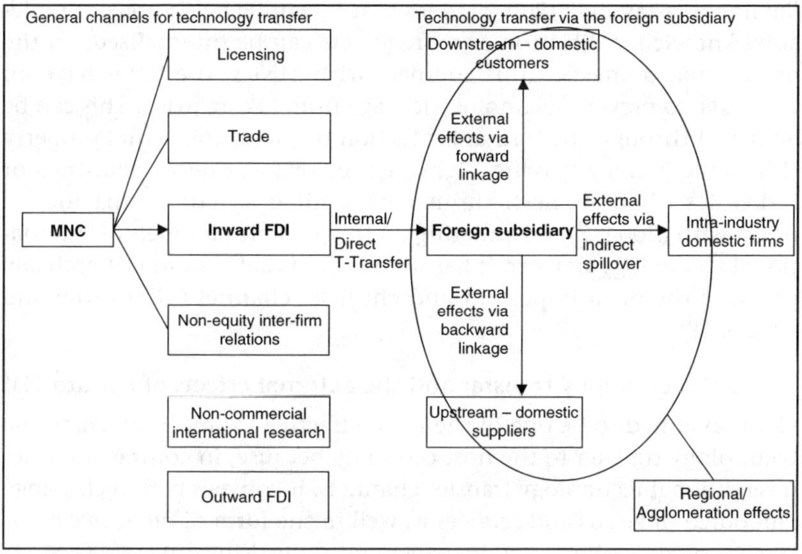

Figure 2.1 General channels of technology transfer and transfer via the foreign subsidiary

Lichtenberg (2001: 490) also propose *outward FDI* as possible channel for knowledge transfer. Moreover, one can also think of non-commercial international scientific research efforts, which serve as a conduit of international knowledge to local companies via domestic research and development (R&D) institutes or universities. When we turn later to empirical studies which aim to assess the importance of inward FDI for technology transfer, it becomes important to check if the channels listed above have also been taken into consideration.

Determinants of the mode of technology transfer

The channel selected for technology transfer depends on the characteristics of the *technology itself* – e.g. its age and complexity – and the features of the *host country* – e.g. the level of education of the workforce, labour skills, technology transfer requirements and competition. Mansfield and Romeo (1980) conclude that the more modern and complex a technology, the more MNCs prefer to transfer it to an affiliate rather than to a third party. Kogut and Zander (1993) show that the more tacit or complex the technology, the more likely it is to be transferred to a fully-owned subsidiary. On the other hand, codifiability and teachability improve the feasibility of licensing.

Smarzynska-Javorcik (2004) argues that inward FDI can prevent undesired knowledge diffusion as more effects can be internalised. To the extent that domestic firms compete with MNCs, the latter have an incentive to prevent technology leakage from taking place. This can be achieved through the formal protection of their intellectual property (IP), trade secrecy, paying higher wages or locating in countries or industries where domestic firms have limited imitative capacities to begin with (2004: 3). Considering all the possible channels of international technology transfer, it has been argued that inward FDI probably provides the most important and cheapest channel (Blomström and Kokko, 1996).

Internal technology transfer and the external effects of inward FDI

FDI is assumed to be one of the most effective forms of international technology transfer to the host economy because, in contrast to other international technology transfer channels, it conveys both technology embodied in goods and services as well in the form of *intangible assets* such as organisational and management skills (Kinoshita, 2000: 5). As Meyer (2003) states, some technology is codified in blue-prints or embodied in machinery and can readily be applied in the host economy. However, the transfer of more complex capabilities – such as skills to assess available technologies and to select the most suitable ones, or the managerial skills to improve organisational arrangements – is more difficult. Such capabilities are often *tacit* and reside not only within the firm, but also in its environment – e.g. in the form of linkages with other firms, or in the educational system (2003: 11). There is also an argument that most MNCs transfer knowledge required for specific application and adaptation, but not the deeper knowledge required for *independent technological innovations* and *strategic decision making* (2004: 12).

There is no unique theory of technology transfer. The theoretical literature rather suggests different classifications and terminologies and attributes varying characteristics to technology transfer types. We follow the definition by Blomström and Kokko (2002), who differentiate between the internal effects of MNC presence, which refers to *direct or internal technology transfer* between the MNC and the foreign subsidiary, and external effects. External effects or *indirect technology transfer* run from the foreign subsidiary to other domestic enterprises. They develop in two dimensions. They run vertically along the supply chain (affecting upstream suppliers through so-called *backward linkages*) and downstream (affecting customers through *forward linkages*). These linkage effects are also described as inter-industry effects, inter-sectoral effects, or vertical spillovers. The second dimension of external effects runs horizontally from the foreign subsidiary to domestic enterprises within the same sector, which are also called intra-industry effects or intra-industry spillovers (see Figure 2.1, right-hand side).

The literature also refers to *inter-firm linkages* between MNCs and domestic firms such as franchising, strategic alliances, subcontracting and modular production (Altenburg, 2000; Majcen, Radošević and Rojec 2003c; Meyer, 2003). In our terminology, equity inter-firm relationships such as franchising are captured by direct technology transfer if it involves foreign ownership of at least 10%. However, non-equity inter-firm relationships belong in the group of external effects, depending on the linkage dimension (vertical, horizontal), given that the MNC has a foreign subsidiary in the host economy. We focus on the nexus between foreign subsidiaries and technology transfer, so non-equity inter-firm relationships between domestic enterprises and MNCs, which do not have a foreign subsidiary in the host economy, constitute a separate channel of international technology transfer (see Figure 2.1).

The benefits of direct technology transfer from the MNC to the foreign subsidiary can be fully internalised in form of efficiency enhancement by the foreign subsidiary and/or the MNC. On the other side, positive indirect technology transfer takes place between the foreign subsidiary and other local companies when the entry or presence of MNC affiliates leads to efficiency enhancement in the host country's local firms and the MNCs do *not* internalise all the benefits (Blomström and Kokko, 1996). This is based on the idea that MNCs are carriers of superior technology, which has certain characteristics of a *public good* and can therefore spill over into the host economy (Markusen, 1995, Caves, 1996, as cited in Görg and Strobl, 2001: 723). Meyer (2003) suggests that spillovers arise from *non-market transactions* when resources, notably knowledge, are spread without a contractual relationship between the owner and the recipient of the knowledge (2003: 4). Kvinge (2004) defines this kind of effect as non-pecuniary or '*pure spillover*'. In our terminology, 'pure spillover effects' can be attributed to intra-industry effects. Kvinge (2004) argues that external effects can also have pecuniary externalities, which exist due to the interdependence among producers through the market mechanism (2004: 9). In this class of externalities we could count the inter-industry linkages and inter-firm relationships which imply some form of contractual relations. Moreover, Kvinge (2004) argues that the presence of MNCs could stimulate the development of industrial clusters and the provision of a previously lacking infrastructure. This spillover type could potentially generate pecuniary and non-pecuniary externalities. In the terminology of this analysis this effect is classed as a 'regional spillover' which combines agglomeration economies with external effects (see Figure 2.1, right-hand side).

From the above introduction it emerges clearly that international technology transfer can takes different avenues. Technology transfer via foreign subsidiaries is just one option and involves different dimensions. We would expect to see internal effects which can be fully appropriated by the foreign subsidiary/MNC and some spillover effects or externalities, which deliver an additional social benefit.

Channels and effects of internal technology transfer via inward FDI

Given the fact that technology transfer starts within the MNC, one has to pay attention in particular to the channels of direct technology transfer to the foreign subsidiary. The MNC potentially transfers to its foreign subsidiary both the technology embodied in, for example, superior machinery and patents or disembodied knowledge such as R&D activities, marketing or organisational skills (Kvinge, 2004: 9). Similarly Meyer (2003) drawing on Child (1993) suggests that foreign investors are a potential source for knowledge at the technical level which includes new specific techniques (methods for quality measurement, scientific and engineering techniques, market research) and systemic-level techniques embracing new procedures requiring integrative learning and coordination (integrated production systems, budgeting systems). MNCs contribute to knowledge transfer through the direct transfer of information and also by stimulating, directly or indirectly, the generation of new knowledge – for instance, by setting the rules and institutions of the local organisation (2003: 12). A major potential element in the direct transfer of technology is the training of local employees at all levels of the organisation, from low-skilled manufacturing operatives through supervisors to technically advanced professionals and top-level managers (Blomström and Kokko, 2002). The training may be provided in formal training courses in the subsidiary or elsewhere in the network of the MNEs, as well as through on-the-job training in close contact with expatriates or trained local staff (Meyer, 2003: 14).

It is important to note that the parent firm can also have a negative impact on the knowledge generation of its affiliate. Depending on the technological level of the subsidiary or the intensity of competition, the MNCs can provide the affiliate with too few technologies, or the wrong kind of technological capabilities. Moreover, the MNC can adopt a strategy which may restrict the production of its affiliate to low-value activities. This in turn can also reduce the scope for technical change and technological learning both within the affiliate and with regard to spillovers to the domestic economy (Damijan *et al.* 2003a: 9–10).

Channels for intra-industry spillovers

Local firms within the same sector as the foreign subsidiary can theoretically benefit from FDI via at least four different channels: demonstration–imitation, competition, labour mobility and the opening up of new export markets. It is, however, difficult to distinguish one from the other since the mechanism of intra-industry technology spillovers from FDI is complex and often interdependent (Kinoshita, 2000: 5).

Demonstration–imitation effect

Demonstration effects work through the direct contact between local agents and an MNC operating at different levels of technology (Kokko, 1992). When foreign affiliates introduce new products, processes and organisational forms, they provide a demonstration of increased efficiency to other local firms. Local firms may also imitate foreign affiliates through reverse engineering, personal contact and industrial espionage (Kokko, 1992; Perez, 1998). Multinational investment may encourage the entry of international trade brokers, accounting firms, consultant companies and other professional service companies, which then may become available to local firms as well (Blalock and Gertler, 2004: 5–6). Prior to the entry of the MNC local entrepreneurs have limited information about the costs and benefits of new methods. With demonstration, information is diffused, uncertainty is reduced and imitation levels increase (Blomström and Kokko, 2000).

Competition effect

The entry of foreign subsidiaries can increase competition within a sector. Greater competitive pressure faced by local firms could stimulate technical change, the introduction of new products and the adoption of new management methods to increase productivity and to defend market share. As a result of the elimination of deadweight loss through increased competition, there could be producer and/or consumer surplus. This kind of external effect, known as a *competition effect*, is most important in industries with relatively low actual and potential competition and high barriers to entry (Kokko, 1992; Perez, 1998). However, the competition effect can also have a negative impact on domestic firms, when foreign firms with superior technology crowd away demand and reduce, at least in the short term, productivity in domestic enterprises, which could potentially force domestic firms to exit. This is also called the *crowding-out effect* or *business-stealing effect* (Aitken and Harrison, 1999). Hence, if the crowding-out effect dominates, than the overall intra-industry spillover is negative; if higher competition triggers mostly efficiency enhancement, the effect is positive. It should be noted that although the competition effect can trigger efficiency enhancements in domestic enterprises it is not a technology transfer channel in the strict sense. However, aş the presence of MNCs is intertwined with both, the competition effect has to be considered.

Labour mobility effect

MNCs build human capital through the training of local employees, which can spill over from foreign affiliates to other firms as skilled labour moves between employers. Employees may leave MNCs to create or join local firms. These spillovers are especially important for firms that lack the technological capabilities and managerial skills to compete in world markets (Kokko, 1992; Perez, 1998). However, MNCs may also hire talent away from local firms, thereby creating a 'brain drain'. Moreover, foreign firms, which often pay higher wages, may raise wages for all firms in competitive labour markets (Aitken, Harrison and Lipsey 1996).

Opening up of new export markets

Altenburg (2000) remarks that MNCs may open up new export markets for local followers that can build on the country-of-origin reputation of the foreign investors and use the same trade channels.

Channels for vertical/inter-industry spillovers: backward linkages

According to Smarzynska-Javorcik (2004), foreign firms often purchase intermediate goods from domestic suppliers due to (for instance) high transportation costs and/or local content requirements. Backward linkages can operate through six main channels:

(1) Increased demand for intermediate products due to multinational entry, which allows local suppliers to benefits from *scale economies*. This interaction generates spillovers if the transferring MNC is not

reaping the full benefits of the transferred technology – for instance by pressure to lower purchasing prices as it may ften do, due to unequal bargaining power (Smarzynska-Javorcik, 2004: 608–10).

(2) Higher requirements regarding product quality and on-time delivery introduced by MNCs, which provide incentives to domestic suppliers to *upgrade their production management or technology* (also Blomström and Kokko, 2003a).

(3) The demand for local imports and quality requirements may induce the MNC to *transfer knowledge directly to the supplier* (Smarzynska-Javorcik, 2004: 608–10).

(4) Indirect knowledge transfer can happen through labour turnover. Meyer (2003) argues that trained employees may move from the foreign investor to its local suppliers or customers, for instance in a process of *outsourcing*. This is more likely than movements to competitors as employment contracts often inhibit moves to direct competitors (2003: 23).

(5) On the other hand MNCs acquiring domestic firms may choose to rely on *foreign suppliers* (including itself) for intermediates or limit local producers to low-value activities. As a result existing supplier–customer relationships are reduced or break down, which limits the scope for spillover effects (Damijan *et al.* 2003a: 10, Smarzynska-Javorcik, 2004: 608–10). The possible increases of competition in the market for inputs via imports can force local firms either to upgrade – for instance by themselves seeking a foreign partner – or to exit (Meyer, 2000).

(6) In international production networks, established suppliers with global reach develop customer-specific know-how and contract specific technologies in close cooperation with the MNC as a core firm. This raises barriers to entry for smaller firms in emerging markets. At the same time if the local supplier succeeds in upgrading its production in cooperation with a key network partner, there is enhanced scope for *backward spillovers* (Birkinshaw and Hagström, 2000, Majcen, Radošević and Rojec 2003c,).

Forward linkages

Vertical spillover effects can also be realised through forward linkages from the MNC/foreign subsidiary to domestic enterprises in the downstream market (i.e. customers):

(1) In particular, MNCs which outsource the distribution of brand name products often make considerable investments in the

performance of their marketing outlets – e.g. automobile dealers, gas stations, restaurant chains, etc. The MNC may rapidly cover extensive markets while minimising risks and investment in distribution channels whilst maintaining homogeneous standards in their downstream activities. The local firm as distributor thus benefits from the use of an established brand name, a proven business concept and often comprehensive training, which reduces the risk of failure. The most important and rapidly expanding mechanism for downstream relationships with local small and medium-sized firms (SMEs) is *franchising* (Altenburg, 2000: 2).

(2) Many transnational producers of machinery, equipment or intermediate goods provide assistance to their customers, which may be local industrial buyers. MNCs provide *training* to the customer's workforce on how to use the acquired machinery or equipment, and provide *repair services*. There is often also information provision on international quality standards and market trends, as well as help in matching domestic firms with potential customers (Altenbary 2002: 2).

(3) The entrance of an MNC can potentially break *supply bottlenecks* in the host economy, which therefore increases the scope for efficiency enhancements in domestic enterprises (Blomström and Kokko, 2003a).

(4) FDI in infrastructure and business services has a direct impact on the productivity of its customers. In industries such as telecommunication, FDI leads to substantial improvement of the services required by businesses; in other cases, such as accounting or IT services, foreign investors provide services previously not available locally (Meyer, 2003: 24).

To summarise, in this section we have established that inward FDI is just one possible channel of international technology transfer. We defined internal and external effects via foreign subsidiaries and introduced explicit channels. It emerged that technology transfer can be both positive and negative. Table 2.1 summarises the explicit channels for each type of technology transfer and differentiates between potential positive and negative effects based on the literature review.

Technology transfer via inward FDI

The body of theoretical literature and a considerable number of empirical studies have suggested possible factors which determine the exis-

Table 2.1 Explicit channels for internal and external effects via foreign subsidiaries

Channels	Technology transfer, MNE to FIE	Intra-industry spillover	Inter-industry linkages: Forward	Inter-industry linkages: Backward
With positive effects	Technical level (e.g superior machinery and equipment)	Competition	Establishment of marketing Outlets	Higher demand for intermediate products, economies of scale
	Systemic level (e.g. integrated production system)	Imitation, demonstration	To industrial buyers of machinery and equipment (technology, training, information, marketing)	Product quality and delivery requirements
	Direct/indirect knowledge stimulation via setting up of rules and institutions	Opening up of new export markets	FDI in business services	Direct technology transfer from FIE to supplier
	Access to international production network		Breaking supply bottlenecks	Participation in international production network
	Training of employees	Labour mobility (turnover)	⟶	(Outsourcing, spin-offs)
	Standardisation of supplier and distribution channels			Increase of import competition

Table 2.1 Explicit channels for internal and external effects via foreign subsidiaries (continued)

Channels	Technology transfer, MNE to FIE	Intra-industry spillover	Inter-industry linkages: Forward	Inter-industry linkages: Backward
With negative effects	Provision of too few/wrong technology (ies)	Crowding-out/Business-stealing effect		
	Restriction to low-value activities	'Brain drain' via labour mobility to FIEs	→	→
	Elimination of FIE	Raising barrier to entry for new entrants		Restriction to low-value intermediates
				Substitution of local inputs by foreign supplies/Internal production within affliate
				Raising barrier to entry for new supplier entrants (international supply chains)

Source: Compiled by author, based on the literature review in the text, the arrows indicate that the respective effect can have repercussions on other DEs as well – e.g. the 'brain drain' effect to the FIE is not restricted to within-sector firms.

tence, the extent as well as the dynamics of internal and external effects via foreign subsidiaries. The majority of the literature identifies four main determinants for technology transfer via inward FDI:

(1) The total number of MNCs present in a country or sector.
(2) The spatial proximity between MNCs and local firms.
(3) The size of the technological gap between local firms and MNCs.
(4) The absorptive capacity of indigenous firms for adopting MNC technology.

More recent literature also points towards:

(5) The importance of the extent of linkages between the foreign subsidiary and the host economy.
(6) The relevance of the MNC strategy, including the objectives and role of its foreign subsidiary.
(7) The impact of institutional and policy factors.

In this section we shed more light on the serelevant determinants, also taking transition-specific conditions into consideration. A rather complex web of determinants thus emerges. Therefore, at the end of the section a general classification of determinants into host country-, industry-, MNC- and subsidiary-specific categories related to the type of technology transfer is developed.

The 'technology gap'

Findlay (1978), who refers back to Veblen (1915) and Gershenkron (1952), hypothesises that regions or countries with a *large initial technological gap* are more likely to benefit from the spillovers of FDI. As a consequence, they may experience stronger growth of total factor productivity (TFP) relative to advanced regions and countries. Empirical models often therefore assume that the potential for internal and external technology transfer increases in a linear manner with the difference in technological levels between domestic and foreign firms. Perez (1997) argues that a relatively wide technological gap can be easily reduced when a foreign presence is modest and slowly growing, or when industrial policies support R&D efforts by indigenous firms. Conversely, left alone to face market competition and a rapidly increasing foreign penetration, even indigenous firms characterised by a relatively narrow technological gap may be driven out of the market by foreigners (1997: 171–2). Due to an outdated capital stock the former planned economies (FPEs) were to a certain extent technologically backward and in need of micro-restructuring; they, could thus take advantage of technology transfer via inward FDI.

Absorptive capacity and the human capital threshold

A caveat to the technology gap hypothesis is that host countries need a minimum of indigenous human capital to be able to benefit from knowledge transfer by MNEs. This argument is linked to the concept of 'absorptive capacity', which has been theoretically developed by Cohen and Levinthal (1989). 'Absorptive capacity' is defined as the firm's ability to recognise valuable new knowledge, integrate it into the firm and use it productively (Lane and Lubatkin, 1998). Keller (1996), Borensztein, De Gregorio and Lee (1998) and Glass and Saggi (2002) argue that absorptive capacity is a function of technology accumulation and human capital in local firms. Additionally, the firm's *organisational structure* and *combinative capabilities* can contribute to its absorptive capacity (Van den Bosch, Volberda and De Boer 1999). Cohen and Levinthal (1989) argue that R&D stimulates innovation but also increases a firm's absorptive capacity. It develops the firms ability to identify, assimilate and exploit outside knowledge, which is likely to increase the incidence of *technology diffusion*. Technology transfer is thus not an automatic consequence of the presence of others' knowledge stock. Domestic firms have to make their own investment in R&D and employee training, and adapt organisational structures that permit innovation (Kinoshita, 2000: 1).

R&D transfer in the MNC

Following from the concept of absorptive capacity, the question arises as to what kind of R&D MNCs transfer to their affiliates. Knowledge production within MNCs can be divided into two main categories: decentralised applied R&D and highly localised basic R&D (Kvinge, 2004). Applied R&D involves product, process or material adaptation to meet governmental requirements and market or production conditions in the host country, which are the main factors underlying the internationalisation of technology. Basic R&D – i.e. science-based fields of activity and the industry's core technologies – appears to be localised and requires a greater intensity of personal interaction. Tacit knowledge therefore tends to be more concentrated. These activities are influenced by the home country's innovation system (universities, research centres) and agglomeration forces (Kvinge, 2004: 19). Recently there have been tendencies towards greater dispersion of industry-specific core technologies due to locally embedded specialisation, MNC global strategies and governmental policy (Cantwell and Santangelo, 1999). However, in general MNCs mostly apply mature technology in foreign affiliates and undertake basic R&D activities at home or in other highly industrialised countries. Moreover, foreign-owned R&D may discourage or crowd out domestic R&D, leaving total R&D unchanged (Driffield, 2001). If competition from domestic firms is weak, foreign affiliates have no reason to import more and newer technology from their parent companies (Perez, 1997). Taking all this into consideration, the overall potential for the transfer of R&D to affiliates in transition countries seems modest and limited to mature technologies and applied research.

Institutional factors and network alignment

According to von Tunzelmann's (2004) theory of network alignment, a firm will typically be involved simultaneously in three types of different networks:

(a) The network of functions carried out by all productive enterprises (technology, production, finance, management, etc.).
(b) The resource network associated with different types of supplying agents.
(c) The geographic network which extends from the local to the global sphere.

In transition countries, all three types of networks are in a process of deep transformation. The interesting underlying argument is that the same extensions of FDI may have different impacts on growth depending on the degree of network alignment.

Tunzelmann (2004) states that the planned economy suffered from a lack of alignment of resource networks (e.g. in the flows of technological resources) and equally a lack of alignment of functional networks (e.g. between technology and production processes). In transition, the formerly weak global impact increased mainly due to the increased level of international trade and FDI. Privatisation boosted the role of local networks which had been rather weak and dependent on national planning. Simultaneously, with an increasing international impact, the collapse of the old national planning system took place and the old R&D institutes and other nationally based R&D operations were left isolated. Tunzelmann (2004) diagnoses a new form of network misalignment in transition countries. Business enterprise R&D and applied research rather than government or higher-education R&D and basic research received the most severe cuts, which limited domestic firms' growth. Many of the old informal networking linkages survived even where they were not pro-developmental. There is a dangerous dependency on FDI in respect of R&D and firm growth due to underdeveloped national policy, which is essential for forming absorptive capacity (2004: 24, 32–3).

For our empirical focus the network alignment approach suggests that transition countries represent a unique unit of analysis concerning the impact of FDI on technology transfer and economic growth in comparison to other emerging countries. Following from this we would expect some common pattern among transition countries in respect to technology transfer. On the other hand, factors such as the privatisation method, the state and alignment of national innovation systems, domestic FDI and R&D policies could explain differences among transition countries.

The motives for FDI

In most MNCs a global strategy exists, which is responsible for generating the motive for engagement in a certain country. This motive is driven by different considerations, which impact on the time, level and intensity of engagement in the host economy. The motives for FDI can be differentiated into resource seeking, market seeking, strategic asset seeking, or efficiency seeking. Alternatively, one can differentiate between export oriented and domestic market oriented FDI. Tunzelmann (2004) argues that FDI into transition countries was dominated by market seeking rather than efficiency seeking. In his view, market seeking FDI runs out of purpose after making the once-for-all gains from entry. The productivity impact can be sharp at first, but then may level off. So long as the principal learning processes are largely confined to the MNCs home country, long-term accumulation of knowledge and formation of absorptive capacity is restricted (2004: 32–3).

Objectives and role of subsidiary and local competitive environment

Leading on from the strategy behind FDI, the foreign subsidiary itself comes into focus. MNCs transfer knowledge to the extent that it serves the subsidiary to achieve its objectives, which in turn depends on the subsidiary's role and the local competitive environment. A subsidiary engaged in R&D or in the application of the latest technology in the production process would naturally receive more knowledge from the parent than a subsidiary manufacturing products at later stages of the product cycle. Moreover, subsidiaries facing competition from technologically advanced local firms may meet that challenge by upgrading technology with additional support from the parent MNC (Blomström, Kokko and Zejan 1994).

From the international business literature has emerged the concept of 'subsidiary development', which focuses on the process through which MNC subsidiaries enhance their resources and capabilities and, in so doing, add increasing value to the MNC as a whole (Birkinshaw and Hood, 1999). The concept refers to the idea that subsidiaries have unique resources and are able to act with considerable autonomy, which may imply that it may be necessary to allocate them different roles within the greater organisation (Bartlett and Ghoshal, 1989). Birkinshaw (1998) describes internally and externally oriented subsidiary initiatives. With externally oriented initiatives subsidiaries seek to identify new customer needs, develop new suppliers or forge new alliance relationships. Internally oriented subsidiary initiatives seek to make the existing set of relationships within the MNC work more efficiently within the confines of the existing network (1998: 356–66). However, the developmental approach has also dangers with regard to possible 'empire building' in subsidiaries, which increases the internal market administration costs (Birkinshaw, 1998) and lowers its efficiency (Mudambi, 1999). Egelhoff, Gorman and McCormick (1998) favour strong headquarter assignments, which still encourages subsidiary initiative. Arguably, there is some debate and conflicting evidence on to whether a head quarter-centred approach or an FIE initiative exerts the greater influence over subsidiary development (Egelhoff, Gorman and McCormick, 1998: 213). But considering the rationale of the concept of 'subsidiary development', with increased autonomy and initiative the subsidiary can enhance the scope or mandate of its activity, which potentially affects the extent of internal and external technology transfer. For example, when a purely production oriented subsidiary upgrades is functions towards marketing or R&D activities, this in turn could increase innovation and absorptive capacity within the subsidiary. Moreover, the degree of autonomy given to affiliates is likely to trigger more integration of local suppliers. Therefore, from an empirical point of view, we should be interested to what extent the autonomy, mandate and initiative of the subsidiary are determinants of technology transfer.

MNC incentives for creating intra-industry and inter-industry effects

Meyer (2003) argues that MNCs are profit maximisers and thus naturally not interested in creating external benefits for others without obtaining a good price for them. Whether foreign investors allow spillover depends on the opportunity cost of sharing knowledge, and the transaction costs of establishing barriers to knowledge flows (2003: 4). Naturally at the intra-industry level the MNC will try to prevent leakage of knowledge, in particular to competitors. However, the MNC is more likely to share general knowledge (for example, about exports markets), as it is less industry-specific and not part of its core capabilities, and its diffusion to local businesses does not endanger its own competitive advantage (Altenburg, 2000). In general it seems more likely that supply chains may be a better conduit for technology transfer than intra-industry spillovers. For example, by transferring technology to local suppliers, MNCs may be able to improve quality and lower the price of non-labour inputs, which benefits all the firms which purchase from these suppliers. The MNC may realise the full benefit of expansion only if the efficiency of supply markets abroad matches or exceeds that of its home manufacturing base. The 'technology gap' between foreign and domestic producers may limit within-sector technology transfer, so MNCs are likely to procure inputs requiring less sophisticated production techniques for which the gap is narrower (Blalock and Gertler, 2004: 2–3, 8).

Determinants of inter-industry effects

Inter-industry effects depend crucially on the extent of linkages between the FIE and the domestic suppliers and customers. Altenburg (2000) argues that the main reason for weak linkages is the lack of efficient domestic firms able to seize new business opportunities related to FDI. Entry barriers for partnership with MNCs differ considerably, mainly according to:

(a) The type of partnership envisaged.
(b) The motives for MNCs seeking partnership with the local firm.
(c) The characteristics of the industrial activity. (Altenburg, 2000: 5–7)

Foreign affiliates making standardised products with mature, non-proprietary technologies have many suppliers to choose from, and it is not necessary to develop special capabilities in any supplier. Where products are specialised and technologically advanced, on the other hand, affiliates tend to prefer in-house production or to retain relationships with a few selected suppliers. MNCs in price-sensitive segments respond more to wage differences than those in markets where innovation and quality are important. The former are generally relatively 'footloose' and less willing to invest in local skills and supplier upgrading (*World Investment Report*, 2001: 137). Market seeking investors tend to develop substantial supplier linkages – for example, due to host country imposed domestic content requirements or generally less demanding markets. Resource seeking and export oriented investors, in comparison, create relatively few linkages, but these linkages with local suppliers are more competitive and sustainable (Altenburg, 2000: 5–7). If most MNCs in a host economy are export oriented and generally do not supply local customers, only technology transfer through backward linkages can be expected (Smarzynska-Javorcik, 2004). Factors impacting on backward linkages include the type of industry, the technological and economies of scale requirements, the length of time the MNC has been operating in the host country, the transaction costs between parent company and affiliates, the MNC's market position and the host country's trade as well as SME policies (Altenburg, 2000: 5–7). Local procurement by foreign affiliates tends to favour acquisitions as a mode of entry, because of the smaller size of the affiliates, declining costs and risks of local sourcing, the need for proximity and the rising cost of internalisation (*World Investment Report*, 2001: 136–8).

FDI penetration and industrial cluster formation

All the empirical papers implicitly assume a linear relationship between the concentration of the foreign presence in a certain sector or industry and the potential for intra-industry as well as inter-industry effects. However, Tunzelmann (2004) can perceive a situation in which MNCs, having established control over a product market, would start to outcompete each other's foreign affiliates. If a sector is dominated by foreign affiliates, not much scope for intra-industry effects is left. Moreover, as a result of a foreign dominated oligopolistic market structure there could be a dangerous dependency on domestic suppliers. It therefore seems more realistic to assume a non-linear relationship between the concentration of FDI and horizontal effects (Damijan *et al.* 2003b). In respect to regional spillovers, a concentration of related industrial activities in the form of an 'industrial cluster' further encourages FDI and intensifies linkage effects (Damijan *et al.*, 2003: 8; Kvinge, 2004: 12). It follows that FDI in any given industry sector which is clustered in a geographically concentrated area ought to transfer

Table 2.2 A classification of determinants for technology transfer via inward FDI

Determinants	Direct/internal technology transfer	Intra-industry spillover	Inter-industry effects: Forward	Inter-industry effects: Backward
Host country-specific				
Technology gap	X	X	X	X
Factor endowment	X	X	X	X
Property rights	X	X		
Policy factors (trade, local content, JV, SME, R&D, FDI, privatisation method)	X	X	X	X
National innovation system (network alignment)	X	X	X	X
Industry-specific				
Type of sector (primary, etc.)	X	(X)	(X)	X
Type of industry	X			X
FDI penetration	/	X	X	X
Speed of FDI penetration		X		(X)
Absortive capacity in domestic firms	/	X	X	X
Technology intensity of sector	X	(X)	(X)	X
Size and ownership characteristics of domestic firms		X	(X)	X
Competitive environment	X	X		X
Capital and scale requirements	X			X
Cluster formation/FDI penetration	X	(X)	X	X
MNC-specific				
Motive for FDI	X			X
Approach to subsidiary	X		(X)	(X)
MNC market orientation	X	(X)	X	X
MNC mode of organisation, international production networks	X		(X)	X
Strategy in respect to R&D	X	(X)		
Technology transfer (adaption and application or innovation)	X	X		
Market position of MNC (e.g. price sensitivity)	X			X
Technology characteristics (age, complexity, tacitness, codifiability)	X			
Non-equity partnerships with local firms and motives			X	X
Cost and risk of sourcing, proximity	X			X

Table 2.2 A classification of determinants for technology transfer via inward FDI (continued)

Determinants	Direct/internal technology transfer	Intra-industry spillover	Inter-industry effects Forward	Inter-industry effects Backward
Subsidiary-specific				
Objectives and mandate	X	(X)	(X)	X
Type of product (e.g. product cycle, standardisation)	X			X
Role of subsidiary (autonomy, internal external initiative, functions)	X		X	X
Size, age and ownership	X		X	X
Mode of establishment (merger and acquisition M&A, greenfield)	X		(X)	X
Type of R&D (applied, basic)	X	X		
Absorptive capacity	X	/	/	/

Notes: X Identified in the literature reviewed.
(X) Not explicitly mentioned by the literature but the determinant is likely to have an effect from the author's point of view.
/ Already covered in another category.

technology more extensively than same-sector FDI which is geographically dispersed. As argued by Aitken and Harrison (1999), there may be reasons to expect that any benefits to domestic firms from foreign investment would be received first by their neighbours before they diffused to other domestic firms. One mechanism through which this may occur is job *mobility*. Job reallocation in transition countries occurs primarily within rather than across regions (Konings, 2000). From this point of view one ought possibly empirically to pay attention to the regional or the local level as a unit of analysis for technology transfer.

Classification of the determinants for technology transfer

From the above discussion, a long list of possible determinants emerges. Given the fact that direct technology transfer is a *sine non qua* condition for external effects via FDI, it seems obvious that the determinants for internal effects are indirectly also a condition for external effects. However, there is a variety of factors, important in particular for linkage effects, which need additional consideration. However, in the literature reviewed the determinants of forward linkages are not very well developed. We classify all the determinants into host country-, industry-, MNC- and subsidiary-specific categories related to direct technology transfer, inter-industry spillover and linkage effects (see Table 2.2). This differentiation reflects, respectivly, micro-, meso- or macro-perspectives of the empirical studies. This overview is meant to serve as a checklist for testing the thoroughness of empirical studies. The task of the empirical analysis is to assess the relative importance of the determinants, which could facilitate conclusions about a possible *hierarchy of conditions* which must be met if positive, negative, or neutral internal and external effects are to happen.

Principal research questions

From the body of theoretical work a number of basic research questions emerge. The following issues are central to our analysis of empirical evidence in the next Chapter.

The first task is to test if internal and external technology transfer are a *real-world phenomenon* or not. This includes the sign of the effects (positive, neutral, negative) and any quantification. When undertaking such analysis, one has to take into account other possible channels of international technology transfer to the host economy, such as international trade. Otherwise it is not possible to show how much of these effects can be attributed to the presence of foreign subsidiaries. Moreover, we are interested in the dynamics of technology transfer – i.e. how does intensity or different channels of technology transfer change over time?

Second, one has to establish the *explicit significant channels* for each type of technology transfer and the respective significant determinants at the host country, industry, MNC and subsidiary level. Otherwise, we have no knowledge about the conditions and mechanisms of technology transfer via foreign subsidiaries. For example, it could be possible that linkage effects are dominated by backward channels and, in particular, higher demand for intermediate products in a certain sector and stage of production. Only this insight could allow policy makers to draw conclusions about targeted public policies – for example, to upgrade local suppliers – if it can be established that local capabilities are the a decisive condition for foreign subsidiaries to choose local sourcing.

Notes

1. Further one can differentiate into fully-owned foreign subsidiaries (100%), majority foreign-owned subsidiaries (50–99%) and partially-owned foreign subsidiaries (10–49%).
2. For example Vernon (1996) proposed that the choice of production location changes with the maturity of a product over its cycle. Knickerbocker (1973) suggested that internalisation is a reaction of MNCs in a oligopolistic markets.

3
Empirical Studies: Approaches, Methodological Problems and Findings

Björn Jindra

Chapter 3 maps out the empirical research in the area of technology transfer via foreign subsidiaries. The bulk of the empirical work has been produced by econometric studies employing the production function approach and quantitative data. After discussing the methodological problems involved, a simple form of meta-analysis is employed to interpret the findings of nineteen studies from transition countries. However, this chapter also considers the newly emerging research from mezzo-analysis, which introduces *qualitative data* into the study. The general aim is to explore the methodological challenges, limitations and insights of the two approaches. This should allow us to develop a more thorough understanding of the current state of empirical research in the field. Moreover, extensions of existing work are suggested where appropriate, and potential new research trends are highlighted.

Methodological challenges

The are at least three possibilities for estimating technology transfer effects using regression analysis of quantitative data: (1) the estimation of the firm's supply curve, (2) the combination of a dominant firm/competitive fringe framework with a model of firm/industry dynamics and (3) the production function approach. In supply curve estimation based on the profit maximisation condition, the supply curve is shifted outward because of cost-reducing or demand increasing spillovers (Levin and Reis, 1988). There are no implicit assumptions about competition. The disadvantage of this approach is that it assumes that firms produce just one type of product and charge on average the same price (Jensen, 2002: 17, 23). In the dynamic competitive fringe framework, firms face uncertainty about their production efficiency and learn about it while operating in the industry (Jovanovic, 1982). It is assumed that domestic firms' production is affected by cumulative technology shocks or technology spillovers (Sun, 2002). This model can be used to derive empirical predictions for the growth and survival of domestic firms. Kosova (2003) argues that this approach has three advantages over the standard production function framework: (1) it does model firm or industry dynamics; (2) it avoids the endogeneity problem, because the model stipulates that the firm's growth depends on firm size, age and market demand; (3) the introduction of cumulative technology shocks permits the incorporation of external technology effects and their separation from the crowding-out effect (2003: 5).

The standard production function approach

However, in the general literature on technology transfer the production function approach is dominant. The production function is a macroeconomic concept, but it is assumed that each firm has its own production function. The presence of inward FDI is treated as an input factor and the impact on a firm's productivity is estimated. Kvinge (2004) argues that the implicit assumption of firms operating on their production frontier (PF) may constitute a problem. Although this may be correct over a longer time horizon, in the short term there may be problems with market conditions, hiring of employees or financing of necessary investments. (2004: 31). Competition is considered to be perfect. Otherwise imperfect competition is taken account of by a mark up factor between price and the value of the marginal productivity of inputs (Barrell and Pain, 1997) or producer concentration is controlled by means of the Herfindahl index (Blomström and Wolff 1994; Kokko 1994; Sjöholm 1999). However, Blalock and Gertler (2004) argue that the production function estimation may confound the productivity gains from technology transfer with the efficiency losses from increased competition[1] (2004:7). In other words the production function approach cannot differentiate between the crowding-out effect and the technology transfer effect. However, both effects are the result of the MNC's entry into the domestic market and could arguably be treated as a net effect.

The impact of the internal and external effects of FDI can be measured indirectly by considering the Solow residual (SR) of output growth as the rate of technological change after subtracting the growth rates of labour and capital. An alternative way is to include the technology variables directly in the production function and to estimate the TFP, a method closer to that of endogenous growth models.[2] This approach provides a way to study the various factors that affect productivity growth, which is done by employing the growth accounting approach and decomposing TFP into factors internal and external to the firm, such as R&D activity, human capital and channels of technology transfer (Damijan *et al.* 2003: 7).

When estimating direct technology transfer, domestic firms and firms with foreign participation can be directly compared in terms of the productivity level and the factors determining productivity. When estimating external effects, a linear relationship between FDI penetration and the potential for technology transfer is assumed. Horizontal effects are usually estimated by taking the stock of FDI in terms of employment, equity, or output of a certain sector as a proxy for the technology transfer potential. If the coefficient for foreign presence is significant for productivity changes in domestic firms within the same sector, this is taken as evidence for horizontal technology transfer (negative or positive). Estimation of intra-industry effects might pick up inter-industry effects, depending on the classification of the sector. Blalock (2001) suggested a way to capture inter-industry effects separately by incorporating a direct requirements coefficient derived from the input–output accounts into the empirical model, which permits estimation of forward and backward linkage effects. The FDI stock in terms of domestic sales/purchases in the supplying/purchasing sector is taken as proxy for the technology transfer potential to domestic firms in the related industries. However, the necessary input-output tables are difficult to obtain. Often the input-output table of one year serves as basis of analysis for a sample stretching over many years, which does not capture the changes of input-output relations over time.

Measurement of variables

The dependent variable is either TFP productivity growth or the growth of output per employee/hour. As a measurement of output, some studies employ gross output while others use value added. It is necessary to deflate input and output to track the changes in quantities rather than prices. However, quality improvements embodied in inputs are not taken into account (Kvinge, 2004: 30–1). There are substantial measurement problems with the factor capital in the former socialist countries, due to poor accounting standards and a tendency to misstate the value of capital. Materials used in production (Damijan *et al.*

2002, 2003a/b), depreciated capital reported in income statements (Kinoshita, 2000) or energy consumption (Djankov and Hoekman, 1998) are used as a proxy for capital utilisation. Technology is an inherently abstract concept. None of the available measures of technology used as explanatory variables – such as R&D expenditures, numbers of new patents, payments for licenes and royalties – cover more than a part of the concept (Blomström and Kokko, 1996: 3). As proxies for the absorptive capacity of the firm, intangible assets, R&D spending or human capital are used. The exclusion of a proxy for R&D leads to the danger of attributing effects to inward FDI rather than to inward FDI undertaking R&D (Kvinge, 2004), which can alternatively be captured by forming interaction terms between FDI and R&D.

Data

Earlier econometric studies researching external effects used cross-section data. More recent studies argue that panel data is a more appropriate method to determine technology transfer effects (Aitken and Harrison, 1999; Görg and Strobl, 2001; Damijan *et al.* 2002, 2003). Cross-section specifications do not permits the identification of the direction of causality between FDI and productivity improvements. Panel data analysis provides techniques to deal with this causality problem, and also allows us to control for firm-specific effects that are time invariant and possibly correlated with a foreign presence in the sector (fixed effect (FE) or random effects (RE) models, see below). Failure to control for such effects may lead to biased results (Konings, 2000: 10; Sinani and Meyer, 2004: 4). Finally, panel data simply gives more information, more variety, many more degrees of freedom and consequently better efficiency of estimators (Greene 1993; Wooldridge 2002).

The estimation technique for panel data analysis

When analysing the impact of different channels of technology transfer on a firm's TFP, a growth model is estimated augmented by the firm's technology structure. Econometric studies offer the choice between an ordinary least squares model (OLS), an RE model or an FE model. Many empirical studies use more than one technique and try different model specifications. Statistical tests are used in order to find the most efficient estimators. However, the main problem in estimating the production function is endogeneity, which can arise in two ways. First, foreign firms may invest in more productive countries, industries, or firms, leading to a *selection bias*. where the most productive local firms receive FDI, one may over-estimate positive productivity related the effects of FDI. Second, endogeneity can also arise due to the *simultaneity* between input factors and productivity. The input factors capital, labour and material inputs are endogenously determined by the firm's past productivity (Griliches and Mairesse, 1995).

Controlling for selection bias

Due to the fact that foreign investment decisions are not randomly distributed, foreign and domestic firms cannot be treated as homogenous units of analysis (Damijan, 2002: 195). In that case, FDI is a limited dependent variable (Greene, 1993). After estimating the OLS, RE and FE models, the econometric analysis typically applies the Heckman (1979)-type two-step procedure[3] in order to control for possible sample selection bias.

Controlling for simultaneity and firm-specific effects

There are different methods for dealing with the problem of simultaneity, although the optimal solution is still an object of debate. The first possibility is to introduce exogenous variables into a first-order autoregressive process, which reduces the bias in the OLS estimator (Damijan *et al.*, 2002). An alternative strategy is to estimate a model in first differences to eliminate the individual-specific fixed effects and to obtain the estimates of effects of FDI on TFP growth (FE or OLS) directly. Another option is suggested by Anderson and Hsiao (1982), who control for simultaneity in a model by differencing and finding instrumental variables which are not correlated with the error term (e.g. in Sinani and Meyer, 2004). A third option is semi-parametric approach, following Olley and Pakes (1996), where the unobserved productivity can be identified from the firms' observable variable input choice (e.g. in Smarzynska-Javorcik, 2004).[4] A fourth option is to estimate a dynamic version of the production function and then correct for endogeneity using the general method of moments (GMM) in a dynamic panel data framework, following Blundell and Bond (1998, 1999). This approach uses a set of lagged levels' and first-differences' instruments (in Konings, 2000; Damijan *et al.* 2003a, 2003b).

Summing up

From the above discussion it emerges that there are a number of possible problems involved in using the production function approach in order to assess the impact on productivity growth of technology transfer via inward FDI:

(1) A macroeconomic model is applied to micro-units.
(2) Assumptions have to be relaxed, in particular with regard to competition.
(3) The production function approach cannot differentiate between the crowding-out effect and the technology transfer effect.
(4) There is a simplifying assumption that there is a linear relationship between foreign presence and external effect.
(5) The choice and quality of data, as well as the estimation technique, are decisive for the quality of results.

It seems paramount to take account of selection bias in the sample and to control for the endogeneity of input factors. This is done using different techniques, which impacts on the comparability of results.

Meta-analysis of traditional econometric studies

In this section, we analyse a sample of econometric studies using quantitative date to research technology transfer in transition countries during the 1990s. The theoretical considerations about the types of technology transfer and respective determinants outlined in Chapter 2, as well as the relevant research questions from Chapter 3, serve as an analytical framework. Bearing the methodological caveats in mind, the interpretation of the studies is conducted with the help of simple form of meta-analysis. The empirical evidence is synthesised in order to test for a possible pattern of technology transfer in transition economies. Each type of technology transfer, its impacts and respective determinants, as well as relevant methodological issues, are discussed. The results are compared to the theoretical literature and, where appropriate, to other international empirical studies. This gives us some insights about the limitations and possible extensions of the respective methods of analysis applied in the empirical research.

The meta-analysis approach

Table 3.1 lists the results of nineteen studies dealing with technology transfer across twelve different transition countries in the period 1992–2001. The studies have been classified according to the country, the data used, the level of analysis, the branches researched, the econometric approach and findings on the type of technology transfer. Table 3.2 summarises the empirical results according to the type of spillover in the respective countries. This rather simple meta-analysis should allow us to detect any patterns in respect to technology transfer in the

Table 3.1 Overview of econometric studies, 1998–2004

Authors	Countries	Year	Data[a]	Branches	Proxy for foreign presence	Dependent variable[b]	Estimation	Robustness	Technology transfer[c]					Comments
										DE	VE	HE	RE	
Djankov and Hoekman (1998)	Czech Republic	1992–7	Panel/ Firms (domestic)	Manfact., retail, industry, finance	Share in sales	Log (y) gross output, TFP growth	Random effects model	Selection bias (Heckman)	CR	+		–		Direct effect, higher foreign subsidiaries above 50 per cent compared to JVs; no impact of absorptive capacity
Kinoshita (2000)	Czech Republic	1993–8	CS, firm (all)	manufact., industry	Share in employm.	VA growth	OLS – fixed effects	Time and sector dummies	CR	–n.s.		n.s.		Positive spillover only for domestic firms with R&D investment (absorptive cap.), positive spillover in oligopolistic sectors of electrical machinery and radio-&TV (also high-innovation R&D), learning effect
Konings (2010)	Bulgaria, Poland, Romania	1993–7	Panel/ Firm (all)	All branches	Share in sales	Log (*Y*) growth of sales	Diff – GMM	Endogeneity of investment, selection bias	PL BG RO	+ –n.s. +n.s.		–n.s. – –	– +n.s. +n.s.	Negative competition effect dominates technology effect, unbalanced panel, many missing observations for PL

Table 3.1 Overview of econometric studies, 1998–2004 (continued)

Authors	Countries	Year	Data[a]	Branches	Proxy for foreign presence	Dependent variable[b]	Estimation	Robustness		Technology transfer[c] DE	VE	HE	RE	Comments
Bosco (2001)	Hungary	1993–7	Panel/ Firms (all)	All branches	Share in sales	Log (Y) growth of sales	Fixed effect	Endogeneity (first differencing of equation)	HU	+		–n.s.		Foreign presence has direct effect of 8- to 16 per cent increase in sales, higher impact with higher foreign ownership share; dominance of crowding-out effect (capturing of market share); high-technology sectors (capital- and R&D-intensive): overall intense competition, crowding-out and no horizontal positive spillover
Zukowska-Gagelmann (2000)	Poland	1993–7	Panel, industry (SO)	Manufact.	Share in output and employm.	Y/L growth (gross output) and VA	OLS	Calculation of standard errors and *t*-values	PL	+		–		Direct effect depends on the 'technology' gap and competition, some productivity convergence and divergence; rejects contagion effect and the 'technology' gap, high negative competition effect

Table 3.1 Overview of econometric studies, 1998–2004 (continued)

Authors	Countries	Year	Data[a]	Branches	Proxy for foreign presence	Dependent variable[b]	Estimation	Robustness	Technology transfer[c] DE	VE	HE	RE	Comments
Evenett and Voicu (2002)	Czech Republic	1994–8	Panel/ Industry level	Manufact., utilities, services	FDI dummy p. year	Log (y) net of spending for materials		Selection pooling, non-random exits	+				Coherent pattern of direct effects only if data is not pooled, selection bias and when non-random exits are controlled for
Kosova (2003)	Czech Republic	1994–2001	Panel/ Industry level	142 industries	Share in sales	Growth in sales revenues	OLS clusters, random/ fixed effects, Tobit, GEE model	Endogeneity, selection bias (industry, growth), pooling	CR +				Competitive fringe framework/ dynamic firm model, foreign expansion has positive effects on growth and survival of domestic firms; crowding-out is a static initial effect (shake out); industries without foreign presence have higher exit rates; sub sample: horizontal spillover significantly positive on growth only for technology advanced industries

Table 3.1 Overview of econometric studies, 1998–2004 (continued)

Authors	Countries	Year	Data[a]	Branches	Proxy for foreign presence	Dependent variable[b]	Estimation	Robustness	Technology transfer[c] DE	VE	HE	RE	Comments
Schoors and van der Tool (2002)	Hungary	1997–8	CS/firm un-balanced	All branches	Share in sales	Y/L growth	OLS	Selection bias (treatment effects model)	HU +	B+ F –	+		Degree of foreign ownership seems not positively related to direct effects; horizontal spillovers are positive only for firms with sufficient absorptive capacity (human capital per worker); the inter-industry effects are larger than the intra-industry effects; openess (share of exports): FDI in closed sectors has no intra-industry effect and strong negative forward effect. In the very open sectors (manufacturing) positive HE and large and positive BL; FDI and trade openness are complements not substitutes, export-driven productivity growth

Table 3.1 Overview of econometric studies, 1998–2004 (continued)

Authors	Countries	Year	Data[a]	Branches	Proxy for foreign presence	Dependent variable[b]	Estimation	Robustness		Technology transfer[c] DE	VE	HE	RE	Comments
Jensen (2002)	Poland	1993–2000	CS/firm	Food industry		Log Y/L	OLS (and supply curve estimation)	Fixed effect on branch level	PL	+		–	+	Indication of positive horizontal spillovers among the geographically confined community of foreign producers but not for domestic producers
Smarzynska-Javorcik and Spatareanu (2003)	Romania	1998–2000	Panel/ Firms unbalanced	105 industries	Share in output	Log (y) turnover	Fixed effects and OLS with Olley and Pakes (1996) corections	Simultaneity (semipara-est. procedure)	RO		–			Olley and Pakes (1996) results: fully-owned subsidiaries generate a positive intra-industry effect and a negative backward effect and vice versa for partially-owned subsidiaries (mode of entry)
									full		+	+		
									par.					

Table 3.1 Overview of econometric studies, 1998–2004 (continued)

Authors	Countries	Year	Data[a]	Branches	Proxy for foreign presence	Dependent variable[b]	Estimation	Robustness	Technology transfer[c]				Comments
										DE VE	HE	RE	
Damijan *et al.* (2003a)	Bulgaria, Czech Rep., Estonia, Hungary, Poland, Romania, Slovakia, Slovenia	1994–8	Panel/ Firm level unbalanced	Manufact.	Share in industry sales and exports	Log (Y) growth of sales	Fixed effect model	Endogeneity, selection bias (Heckman)	BG	+ n.s	–n.s.		HU SK, poor data quality; majority ownership is n.s. for direct effect; mixed results for the impact of R&D spending and horizontal spillovers (pos. in Slovakia, neg. in ES, HU, LT), exports and imports are only in SL a pos., significant channel of T-transfer
									CR	+	–n.s.		
									ES	+	+		
									HU	–n.s.	–n.s.		
									PL	+	–n.s.		
									RO	–n.s.	–n.s.		
									SL	+	–n.s.		
									SK	+	–n.s.		

Table 3.1 Overview of econometric studies, 1998–2004 (continued)

Authors	Countries	Year	Data[a]	Branches	Proxy for foreign presence	Dependent variable[b]	Estimation	Robustness	Technology transfer[c]	DE	VE	HE	RE	Comments
Damijan *et al.* (2003b)	Bulgaria, Czech Rep., Estonia, Hungary, Latvia, Poland, Romania, Slovakia, Slovenia	1994–9	Panel/ Firms (all), balanced	Manufact.	Share in output adjusted for trade	Log (Y) growth of sales	Sys GMM/ dynamic panel approach	Endogeneity (GMM), Selection bias (Heckman)	BG	–n.s.	–n.s.	+n.s.		In LT and RO direct effects are pos. sig. only for majority-owned FIEs; only backward linkages were tested; direct effects are 50 larger than backward effects and 500 times-larger than horizontal effects, innovative R&D increases only in RO horizontal spillover, whereas it is an obstacle in LV and EE (poor R&D data)
									CR	–	+	+		
									ES	+	–n.s.	–n.s.		
									HU	+	–n.s.	+n.s.		
									LT	–	+n.s.	–n.s.		
									LV	–n.s.	+n.s.	–n.s.		
									PL	–	+	+		
									RO	–	+n.s.	+		
									SK	+n.s.	+n.s.	+		
									SL	+	+	+n.s.		

Table 3.1 Overview of econometric studies, 1998–2004 (continued)

Authors	Countries	Year	Data[a]	Branches	Proxy for foreign presence	Dependent variable[b]	Estimation	Robustness	Technology transfer[c] DE	VE	HE	RE	Comments
Smarzynska (2003)	Lithuania	1996–2000	Panel/ Firm (all), unbalanced	Manufact.	Share in equity weighted by output	Log (y) sales adjusted for inventory	semiparametric estimation	Omitted variable (differencing) selection bias (Maddala two-step procedure) endogeneity semiparametric est.	LT +	+	n.s.		Backward linkages: not significant with Olley and Pakes (1996) corrections at national level, however sig. pos. at regional level and higher for domestic firms, less robust results when MNC is located in other +region; backward linkages are stronger for domestic market oriented rather than export oriented affiliates, the extent of foreign ownership (fully or JV) does not affect the extent of backward linkages; horizontal effects never significant;

Table 3.1 Overview of econometric studies, 1998–2004 (continued)

Authors	Countries	Year	Data[a]	Branches	Proxy for foreign presence	Dependent variable[b]	Estimation	Robustness	Technology transfer[c]	DE	VE	HE	RE	Comments
Damijan and Knell (2003)	Estonia, Slovenia	1994–9	Panel/ Firm (all)		HE: share in industry output adjusted by export of FIEs	Log (y) TFP growth	OLS first differenced model	Selection bias (Heckman)	ES SL	+ +	–n.s. –n.s.	–n.s. –n.s.		No impact of majority ownership on DE; VE results are the same for forward and backward linkages; there is sig. pos. effect for R&D accumulation in Estonian Domestic enterprise; no pos. sig. effects of absorptive capacity and external effects; for Slovenian domestic firms exports and imports are sig. pos. with T-transfer, not in Estonia (impact of privatisation method)

Table 3.1 Overview of econometric studies, 1998–2004 (continued)

Authors	Countries	Year	Data[a]	Branches	Proxy for foreign presence	Dependent variable[b]	Estimation	Robustness	Technology transfer[c] DE	VE	HE	RE	Comments
Vahter (2004)	Estonia, Slovenia	1994/6 2000/1	panel/ firms (all), balanced	Manufact.	Share in assets (adjusted for FIE's own assrts at sector level)	Log (y) sales p. employee	Random and fixed effects model	Selection bias (Hausman two-step), endogeneity (RE)	ES + exp– SL + exp+n.s.	ES DE FIE SL DE FIE	+n.s. + ... + –n.s.		DE: in Estonia the higher FIE productivity is explained by domestic market oriented FDI, whereas in Slovenia the contribution from export FDI is likely to be higher; HE are differentiated into domestic enetrprises and foreign enterprises (FIE), results are explained by comparative advantage and FDI motives

Table 3.1 Overview of econometric studies, 1998–2004 (continued)

Authors	Countries	Year	Data[a]	Branches	Proxy for foreign presence	Dependent variable[b]	Estimation	Robustness	Technology transfer[c] DE	VE	HE	RE	Comments
Lutz and Talavera (2003)	Ukraine	1998–9	Panel/ Firm	Manufact. /7 industries	FDI dummy	Log (Y/I) and Log export volumes	Random effects model/GLS	Endogeneity (two-stage procedure)	UK +		+		Pos. direct effect on labour productivity and export volumes, direct effection labour productivity according to industry; horizontal effects to des are relatively small
Bessenova, Kozlor and Yndaeva (2003)	Russia	1995–2001	Panel/ Firm	83 industries	Share of assets	Log (y) in value added	Random effects and instrumental variable RE	Possible selection bias due to non-random exits and non-reporting	RU	B + F –	+		Horizontal effect n.s. after 1998, the backward linkages in sub-sample of import competing suppliers were negative until 1998 and became pos. thereafter (devaluation)

Table 3.1 Overview of econometric studies, 1998–2004 (continued)

Authors	Countries	Year	Data[a]	Branches	Proxy for foreign presence	Dependent variable[b]	Estimation	Robustness	Technology transfer[c]				Comments
									DE	VE	HE	RE	
Toriak (2004)	Bulgaria, Czech Rep., Hungary Poland Romania	1993–9	Panel/ Plant-level unbalanced		Direct share in capital; hor.: share in total sector output	Log (y) growth in sales	OLS		CR	+	–	–	Test for horizontal effects at regional level (NUTS-2) controlling for agglomeration effects (no. of firms); does not take into equation vertical linkages nor trade effects; does not control for selection bias and endogeneity
									PL	+	n.s.	+	
									BG	+	–n.s.	–n.s.	
									RO	+	–	–	
									HU	+	+n.s.	–n.s.	

Table 3.1 Overview of econometric studies, 1998–2004 (continued)

									Technology transfer[c]				
Authors	Countries	Year	Data[a]	Branches	Proxy for foreign presence	Dependent variable[b]	Estimation	Robustness	DE	VE	HE	RE	Comments
Sinani and Meyer (2004)	Estonia	1994–9	Panel/ Firms (all)	Manufact., equipi- ment, trade	Share of E. equity and sales	Log (y) growth of sales	Fixed effects, GLS panel	Endogeneity (inputs), competition, omitted variables ES	+		+		HE: are positive independent from human capital and investment (lack of absorptive capacity); magnitude of spillover for labour-sales-intensive FDI higher than for equity-intensive FDI; smaller, non-exporting and outsider-owned firms benefits a slightly more, no convergence in most industries

Notes: CS = Cross-sectional data, Panel = panel data.
[b] TFP = Total factor productivity, VA-Value added, Y/L = Labour productivity.
[c] DE = Direct/Internal effects, VE = Vertical effects (forward, backward), HE = Horizontal/intra-industry effects, RE = Regional spillover effects.
+, pos. = positive significant, –, neg. = negative significant, n.s. not significant, sig. = significant.

JV = Joint venture	SO = ???	p = ???	BG = Bulgaria	LV = Latvia
B = backward	EE = ???	GEE = ???	CR = Czech Republic	PL = Poland
F = forward]	T-transfer = Technology transfer	BL = ???	ES = Estonia	RO = Romania
		GLS = ???	HU = Hungary	RU = Russia
			LT = Lithuania	SK = Slovakia
				SL = Slovenia
				UN = Ukraine.

Table 3.2 Summary of the meta-analysis

	Bulgaria	Czech Rep.	Estonia	Hungary	Latvia	Lithuania	Poland	Romania	Russia	Slovakia	Slovenia	Ukraine	No. of Test results	(%)
Direct effects from MNC to foreign subsidiary													*39*	
No. of studies	15													
Positive sign.	*	****	*****	****	*		*****	**			***	*	26	**67**
Negative sign.		*			*		*	*					4	10
No sign.	***	*		*		*		*		**			9	**23**
Intra-industry effects from the subsidiary to local firms													*42*	
No. of studies	18													
+		**	*	*			*	***	*	*	*	*	12	29
–		**					*	***					6	14
n.s.	****	**	****	****	**	*	***			*	***		24	**57**
Backward effects from subsidiary to local suppliers													*17*	
No. of studies	6													
+		*		*	*		*	*	*		*		7	41
–								*					1	6
n.s.	*		**	*	*	*		*		*	*		9	53
Forward effects from subsidary to local customers													*4*	
No. of studies	3													
+														
–				*					*				2	50
n.s.			*								*		2	50

Table 3.2 Summary of the meta-analysis (continued)

	Bulgaria	Czech Rep.	Estonia	Hungary	Latvia	Lithuania	Poland	Romania	Russia	Slovakia	Slovenia	Ukraine	No. of Test results	(%)
Regional spillovers													*10*	
No. of studies	4													
+					*			*					2	20
–		*					**	*					4	40
n.s.	**			*				*					4	40
Total no. of tests results														
Per country	11	14	13	14	7	3	14	16	3	5	10	2	*112*	

Note: **Each star indicates an econometric test result, which is classified as positive/negative significant or as not significant.**

transition countries. The sample is dominated by results for the Czech Republic, Hungary, Poland and Romania whereas Lithuania, Russia and Ukraine are under-represented. The results are strongly influenced by two studies – Damijan *et al.* (2003a, 2003b) – which is an advantage because both are based on large cross-country data sets and employ a unified methodology.

Direct technology transfer

Over two-thirds of the results indicate significant positive productivity effects in foreign subsidiaries from direct technology transfer. This means that foreign subsidiaries did grow faster in terms of productivity in comparison to domestic enterprises. All negative results stem from one study (Damijan *et al.* 2003b). Certainly the negative results for the Czech Republic and Poland are at odds with the consistently positive results for the respective countries from several other studies. Looking at the pattern across transition countries, in regard to direct effects one could cautiously argue that there is substantial positive evidence for the Czech Republic, Estonia, Hungary, Poland and Slovenia. On the other hand, evidence for Bulgaria, Romania and Slovakia is rather mixed.

The direct effect is consistently more often positive than any external effect, and also larger in its magnitude. In Damijan *et al.* (2003b) the direct effects of FDI are found to provide on average an impact on firm's productivity that is larger by a factor of 50 than the impact of backward linkages and by factor of 500 larger than the impact of horizontal spillovers. It is therefore reasonable to conclude that direct technology transfer is the most important channel for technology transfer via foreign subsidiaries in transition economies. Several studies quantify the direct productivity effect. For example, Bosco (2001) calculates that the foreign presence increases sales by 8–16 per cent for firms in Hungary. Evenett and Voicu (2002) find that the foreign presence increases productivity in foreign subsidiaries by over 43 per cent in comparison to domestic enterprises[5] in the Czech Republic. Are the above findings a result of selection bias? Djankov and Hoekman (1998) as well as Evenett and Voicu (2002) claim that foreign investors tend to acquire shares in the largest and most successful domestic firms. In contrast Damijan *et al.* (2003b) find that foreign investors across ten countries undertake more capital- and human skill-intensive investments rather than investments in high-productivity and large-sized firms. Hence, although selection bias exists, its form varies. However, this is not a concern about the validity of results, because apart from the studies by Bosco (2001), Bessenova (2003), and Torlak (2004), selection bias in the data has been controlled for in all studies considered in this sample. It follows that the positive results for direct effects are not explained by the biased choice of foreign investors towards more productive industries and firms. The finding that FDI significantly stimulates productivity growth in foreign subsidiaries should not come as any surprise. It simply proves that foreign subsidiaries can exploit some sort of firm-specific advantage and domestic firms operate at a lower technological level. In other words a technology gap exists between foreign and domestic firms in the transition economies. The dominance of direct technology transfer is supported by ample evidence on positive direct technology transfer from a wide range of international studies for developed as well as developing countries (e.g. Haddad and Harrison, 1993; Blomström *et al.* 1994; Aitken and Harrison, 1999; Blomström and Sjöholm, 1999; Barry *et al.* 2001; Blalock 2001; Girma and Wakelin, 2001; Alverez, Damijan and Knell, 2002, etc.).

Determinants of direct technology transfer

Djankov and Hoekman (1998) find a positive relation between the extent of foreign ownership and the direct technology transfer. In contrast, Schoors and van der. Tool (2002) and Damijan *et al.* (2003a) find no impact of the degree of foreign ownership. Damijan *et al.* (2003b) come to the same conclusion apart from the results for Lithuania and Romania, where only majority foreign-owned enterprise show significant positive productivity effects.

Damijan and Knell (2003) test explicitly for direct technology transfer via trade apart from inward FDI effects. They find that domestic Slovenian firms were able to benefit from technology transfer via imports as well as exports, whereas Estonian firms show no significant effects from trade. They link this phenomenon to the method of *privatisation,* which involved less FDI in Slovenia and which led to a development of an arm's length relation to MNCs. It can therefore be argued that the interplay between FDI market orientation and the privatisation method impact on the relevant international technology transfer channel and the extent of direct technology transfer via foreign subsidiaries.

Torlak (2004) finds productivity growth in foreign subsidiaries much more pronounced in firms in Bulgaria, Poland and Romania comparison to the Czech Republic and Hungary. He argues cautiously that this might be explained by a low initial productivity level in firms receiving foreign capital in the first three countries. This would give support to the 'technology gap' hypothesis. Following from this, we would expect across our sample of studies the strongest direct effects in more backward countries such as Bulgaria or Romania. However, as outlined above, the more advanced countries such as the Czech Republic, Hungary and Poland show more consistently positive evidence. Unfortunately, other cross-country studies do not account for the impact of the 'technology gap'. As a result, we cannot draw any final conclusions about the validity of the linear relationship between the direct effects of the 'technology gap' and technology transfer.

Generally, most studies find that the extent of the direct technology transfer differs according to the industry or sector, which is observed via the use of dummies. The effects are not explicitly differentiated according to certain industries/sectors, apart from the study by Evenett and Voicu (2002). They find for the Czech Republic in nine out of ten industries positive significant direct effects (including transport equipment, textile, primary metals, etc.) and no significant effects for the service industry publishing.

Findings in the light of the theory

Lets as try to put the above outlined empirical findings in regard to direct technology transfer into a theoretical context. There is strong evidence for direct productivity-enhancing effects via inward FDI. However, the econometric studies do not contribute to showing the explicit channels at work – i.e. technical changes, systemic changes, access to international production networks, or standardisation of suppliers and customers. As a result, we know that direct technology transfer is a real-world phenomenon, but we can not explain how it takes place. There are indications that trade and FDI are substitutes for international technology transfer.

The list of determinants tested for direct technology transfer via inward FDI is somehow limited, because only the privatisation method, the degree of foreign ownership and the trade propensity of FDI have been considered. It is not clear to what extent technology transferred across countries or industries is a linear function of the 'technological gap'. Evidence is missing about the impact of determinants such as factor endowment, competition, technology intensity of industries, MNC-specific factors (e.g. technology characteristics, strategy, mode of organisation, etc.) as well as the whole range of foreign subsidiary-specific factors (role, objectives, mandate, type of R&D, absorptive capacity, etc.).

Horizontal/intra-industry effects

If we have evidence that a positive direct technology transfer takes place, it follows that are need to question to what extent technology spills over from foreign subsidiaries to domestic firms operating in the same sector. Most studies focus on this technology transfer type in their research. However, the evidence is not clear-cut, because by far the majority of estimations produce no significant results. Still in twelve out of forty test estimation results horizontal effects were positive and in six instances negative. The positive and negative results are evenly distributed across different studies. Looking at the range of detected horizontal effects per country does not offer a clear-cut pattern. On the contrary, the results are fairly contradictory particularly for the Czech Republic and Romania (see Table 3.2). Where positive horizontal effects were detected, their impact was smaller in comparison to inter-industry impacts (Schoors and van der. Tool, 2002; Damijan *et al.* 2003b). The studies in the sample do not provide any information about the explicit channels for horizontal technology transfer.

The dynamics of crowding-out and technology transfer effects

What are the dynamics of the crowding-out effect and the technology transfer effect, which are both results of the entry of MNCs? Most previous studies include only a single measure for FDI presence. Therefore, they cannot differentiate between crowding-out and technology transfer effects and estimate only a net impact.[6] Studies which detect negative spillover effects at the intra-industry level generally argue that the crowding-out effect dominates the positive technology transfer effect (Zukowska-Gagelmann, 2001; Jensen, 2002). However, if the net effect is zero one does not know whether the two effects are small and unimportant, or both large. A study by Kosova (2003) takes a new tack by estimating the survival and exit rates of domestic firms rather than estimating a production function. Kosova finds that foreign entry increases the exit rates of domestic firms in the short term (the shake-out phase), but subsequently the growth of the foreign industry segment is accompanied by increases in both the growth rate and the survival of domestic firms. Hence, crowding-out is a short-term or static effect, whereas the technology transfer effect develops over time.[7] The studies employing the production function approach cannot shed sufficient light on the dynamics of the crowding-output/technology transfer effects. However, Torlak (2004) argues for Hungary that the MNC entry forced less productive Hungarian firms to exit the market, before they had managed to developed a knowledge base. Horizontal spillovers are absent, and a large 'productivity and technology gap' exists between domestic and foreign enterprises. It thus seems that the dynamics of the impact of FDI on productivity are characterised by the immediate and direct market stealing effect and a long-term and conditional learning process (2004: 18).

The determinants of horizontal effects

Various studies have dealt with the technology intensity of the industry/sector, and competition and technology as determinants. Kosova (2003) finds positive horizontal effects only for technologically advanced industries in the Czech Republic. Kinoshita (2000) finds positive horizontal effects only for oligopolistic markets which also showed innovation R&D intensity (electrical machinery, radio and TV) in the Czech Republic. However, Bosco (2001) finds for Hungarian high-technology industries, which were capital- and R&D-intensive, an overall intense competition, crowding out and no positive horizontal effects. Zukowska-Gagelmann (2001) finds negative horizontal spillovers for Poland,[8] higher in highly competitive industries with a low 'technology gap'. Surprisingly, no negative spillovers were detected at all for 'high-technology gap' industries. Under low competition conditions the negative impact is less detrimental (2001: 155–61). Despite the fact that there is consistently a 'productivity/technology gap' between foreign and domestic firms, the 'technology gap' hypothesis has not been systematically analysed in respect to horizontal spillovers at the national or industry level (apart from Zukowska-Gagelman, 2001). However, there is some evidence from international studies that the linear relationship between technology transfer effects and the 'technology gap' does not hold (Haddad and Harrison, 1993; Kokko, 1994; Kokko, Blomström and Kokko 1996).

A number of studies try to include the concept of absorptive capacity when analysing intra-industry effects. Damijan and Knell (2003) used intangible assets as a proxy for Estonia and Slovenia and do not find that absorptive capacity has an impact on horizontal spillovers. Sinani and Meyer (2004) find positive horizontal spillovers independent from human capital and investment, which is interpreted as a general lack of absorptive capacity among the Estonian firms. On the other hand Kinoshita (2000) finds positive significant horizontal spillovers for Czech firms only when they invested in R&D. Following Cohen and Levinthal (1989) she differentiates the effects of R&D into a learning effect and an innovation effect and the former of higher importance for horizontal spillovers. Also Schoors and van der. Tool (2002) find for Hungary that horizontal effects are significantly positive only if the domestic firms have sufficient investment in human capital. In other words labour- as opposed to skill-intensive firms are less likely to benefit from the presence of MNCs within the sector. The two large cross-country studies by Damijan *et al.* (2003a, 2003b) show mixed results for the impact of R&D on the extent of horizontal spillovers. This is most likely due to the fact that intangible assets make a poor proxy for R&D. It could be concluded that absorptive capacity in relation to factor endowment potentially plays a role in accordance with the theoretical literature. However, the results depend very much on the proxy chosen.

Sinani and Meyer (2004) detect for Estonian firms positive horizontal spillovers for all ownership groups; however, the magnitude of the coefficient is significantly larger for outsider-owned firms than for state-owned and insider-owned firms (2004: 17). This finding is in line with the literature which stresses the importance of outsider ownership for restructuring (Frydman *et al.* 1997; Djankov and Murrell, 2002). Interestingly Smarzynska-Javorcik and Spatareanu (2003) find for Romania that only fully-owned subsidiaries generate positive horizontal effects, whereas partially-owned subsidiaries create negative effects. This potentially could be linked to the finding by Damijan *et al.* (2003b), where in Romania only majority-owned subsidiaries experienced positive direct productivity effects. In sum, one could argue that the extent of foreign ownership and corporate control has an impact on horizontal effects.

Bessenova, Kozlov and Yudaeva (2003) and Damijan and Knell (2003) also take account of trade as a possible source for technology transfer to domestic enterprises. Damijan and Knell (2003) find that domestic Slovenian firms were able to benefit from technology transfer via imports as well as exports, whereas Estonian firms show no significant effects from trade. This is linked to different privatisation methods in both countries. This argument could suggest that FDI and trade are substitutes for technology transfer. However, Bessenova Kozlov and Yudaeva (2003) finds that Russian domestic firms benefit from both import liberalisation and inward FDI. Schoors and van der. Tool (2002) find for Hungary that FDI in relatively closed industries (low share of exports) has no horizontal effects; however, very open sectors (manufacturing) show positive intra-industry effects. It therefore seems that FDI and trade are complements rather than substitutes for technology transfer.

Findings in the light of theory

Given the importance of horizontal spillovers in the theoretical literature, the empirical evidence is sobering. The sample shows by and large contradictory estimation results for each country. The studies in the sample do not provide any information about the explicit channels for horizontal technology transfer. In terms of dynamics, there seems to be an initial strong and negative crowding-out effect, followed by an increasing but conditional technology transfer effect. In respect to determinants, empirical studies do not explicitly take account of the 'technology gap' hypothesis. The evidence for absorptive capacity seems to have some substance, but results vary according to the quality of the proxy. Technology intensity seems to impact positively on the extent of intra-industry effects whereas, as expected, a highly competitive environment seems generates crowding-out effects. The degree of ownership seems to be relevant for the foreign subsidiary as well as the domestic firm in regard to realised horizontal spillovers. The privatisation method or broader the stance towards FDI has an impact on the dominance of inward FDI or trade as international technology channel; however, both channels seem to be complementary. Factor endowment, FDI market orientation and the motive for FDI seem to impact on the presence and extent of horizontal spillovers. Unfortunately, important host country-specific factors such as the state of the national innovation system, SME and R&D policies, have not been considered.

Why are horizontal effects absent?

We remain with the finding that horizontal effects tend not to be detected in transition countries, which is supported by other international studies for developed and developing countries (Görg and Greenaway, 2001; Kvinge, 2004). Görg and Greenaway (2001) list different possible reasons: (1) MNCs might be very effective in protecting their technology advantages; (2) studies have been carried out at the aggregate/sectoral level or use cross-section analysis. However, in our meta-analysis only three out of nineteen studies use cross-sectional data and apart from two studies (Zukowska-Gagelmann, 2001; Kosova, 2003) all studies use firm-level data. Damijan *et al.* (2003b) suggest other explanations such as the poor quality of data, short panels and inappropriate econometric techniques (2003b: 2). The poor quality of data can not be excluded as a reason; however, the findings in respect to horizontal spillovers were similar using different data sources and using a small or large data-set. Indeed, some of the panels are relatively short (Schoors and van der. Tool, 2002; Lutz and Talavera, 2003; Smarzynska-Javorcik and Spatareanu, 2003) but the majority of studies work with panels extending over five years. With regard to econometric techniques, the assessment is difficult. However, taking the example of the Czech Republic it emerges that even studies controlling for endogeneity, sample attrition and selection bias, although in different ways, still produced contradictory evidence (Djankov and Hoekman, 1998; Kinoshita, 2000; Kosova, 2002; Damijan 2003b; Damijan *et al.* 2003a). It is beyond the scope of this analysis to test which econometric technique is the most appropriate.

Yet another point made by Damijan *et al.* (2003b) is that all studies assume a *linear relationship* between foreign penetration of the respective industry and the productivity growth of local firms in that industry. They instead make a case in favour of an inverted U-shaped relationship[9] (2003b: 2). If there is indeed a non-linear relationship, then the effects cannot be captured by studies employing linear estimation. Depending on the time horizon, studies for the same sector in one country would produce contradictory results, which is indeed the case in our sample. In order to understand the intra-industry effects properly we should take the findings by Kosova (2003) and Torlak (2004) into consideration. It seems paramount to differentiate between crowding-out and technology transfer effects. Following Kosova (2003) we can estimate the exit/survival rates of domestic enterprises to determine the initial crowding-out phase. This is followed by an production function estimation for the subsequent time period, which should measure the actual horizontal technology transfer effect. In addition, we need to account for possible determinants of particular absorptive capacity. If this combination of methods does not produce significant results, we would be left with three possible explanation for the absence of horizontal effects:

(a) MNCs are effective in preventing technology leakage within the same sector (which could be related to oligopolistic market structure).
(b) Domestic enterprise lacks the absorptive capacity to benefit from technology transfer (or, more generally, lacks the means to compete).
(c) Horizontal technology spillovers via foreign subsidiaries is a purely theoretical concept.

Vertical/inter-industry effects

With the exception of two papers, international studies in general do not take account separately of backward and forward linkages as channels for technology transfer. Kugler (2001) finds positive inter-industry effects in Columbia. Blalock and Gertler (2004) find positive backward linkage effects in Indonesia. Studies for transition countries seem in that respect closer to the research frontier, as we find five studies in our sample analysing backward linkages and two studies considering forward linkages.

Schoors and van der. Tool (2002) find positive backward linkages and negative forward linkages for Hungary. Moreover, in industries relatively protected from trade[10] they find strong negative forward effects, and in very open industries backward linkages are particularly large and positive. Smarzynska-Javorcik and Spatareanu (2003) differentiate in respect to ownership and find negative backward linkage effects from fully-owned foreign subsidiaries and positive backward effects from partially foreign-owned subsidiaries in Romania. This could be treated as evidence that JVs/minority foreign-owned firms tend to source locally. However, according to Smarzynska-Javorcik (2004) in Lithuania the extent of foreign ownership is found not to be a relevant determinant of backward linkages. She finds positive backward linkages not at the national but at the regional level. Moreover, backward linkages are stronger for domestic market oriented vs. export oriented FDI. Damijan *et al.* (2003b) find positive backward linkages in the Czech Republic, Poland and Slovenia. In Bulgaria, only foreign affiliates can attain these backward spillovers, while in Lithuania and Latvia even negative backward spillovers for foreign affiliates are detected. The latter finding may well be caused by the very poor coverage of firms in the sample. Still, where positive backward linkages exist they are on average ten times larger than horizontal effects. Surprisingly, the absorptive capacity of firms does not seem to have an effect on backward linkages. However, this result again could be attributed to intangible assets being a poor proxy (2003b: 18). Damijan and Knell (2003) also find no significant backward or forward linkages in Estonia and Slovenia after taking absorptive capacity into the equation. Finally, Bessenova, Kosova and Yudaeva (2003) find significant positive backward linkages and negative forward linkages for Russia. Interestingly they split the sample into the period before and after the Russian crisis. As a result, they find backward linkages negative until the crisis and positive thereafter, which can be explained by the effects of devaluation and the substitution of imports by local supplies. This finding hints first at non-linearity with regard to backward linkages[11] in the sense of structural breaks, and secondly at the exchange rate as the determinant of the extent of backward linkages. The absence of forward linkage effects could potentially be explained by the export orientation of FDI or the limited integration of FDI into the domestic economy, by concentration of FDI towards the end of value chains, or a 'technology gap' between the foreign subsidiary and domestic producers in the downstream market.

In sum, one could cautiously argue, having in mind the limited number of studies, that there are indications that backward linkages could be the second most important channel for technology transfer via foreign subsidiaries. Backward linkages seem to be the strongest positive channel with regard to external effects. The limited evidence available points towards a negative impact for forward linkages. Vertical linkages are subject to the mode of entry, the degree of foreign ownership, the market orientation of FDI and exchange rate movements. It can be assumed that the explicit channel for backward technology transfer is increased demand for intermediate products. To what extent technology is transferred directly or indirectly by setting product and quality requirements remains unknown. Moreover, in cases where backward linkages are negative, the causes cannot yet be explained. Possibly suppliers are restricted to low-value activities or have lost demand due to substitution by imports or in-house production within the MNC. At this stage of research qualitative findings could be supplementary to the quantitative results. In particular for backward linkages, the subsidiary- and MNC-specific determinants could play a role but have not been subject to systematic analysis in a sample of econometric studies. These two types of technology transfer channels via inward FDI should be a source of further research.

Regional effects

Only a few international quantitative papers depart from using countries as the level of analysis in order to detect the relevance of regional effects (Kvinge, 2004). Aitken and Harrison (1999) in Venezuela and Haskel, Pereira and Slaughter (2001) in the United Kingoom, test for the possibility that external effects pertain to a local region smaller than the host economy, but they find no evidence to support this claim. In contrast Zhiqiang (2002) finds positive evidence of spillovers at regional level within China, as does Sjöholm (1999) in Indonesia.

In our CEE-specific sample, four studies consider regional effects. Konings (2000) measures horizontal effect at a regional level.[12] The results for horizontal effects are not significant at the national as well as the regional level for Romania and Bulgaria. However, Konings detects negative regional spillovers for Poland for which horizontal effects at the national level were not significant.[13] Jensen (2002) finds negative horizontal spillovers at the national level and positive horizontal effects only among the geographically confined community of foreign producers, but not for domestic producers. Smarzynska-Javorcik (2004) finds significant positive backward linkage effects only at the regional level, not at the national level. Torlak (2004) tests for horizontal effects at the regional level and finds large and significant positive regional spillovers.[14] In order to control for location-specific variations in productivity due to agglomeration economies or other region-specific effects Torlak introduces *firm density* as an additional variable. As a result, the coefficient for FDI presence goes down and he finds negative regional effects for the Czech Republic and Romania, a positive effect for Poland, and not significant results for Bulgaria and Hungary. These findings indicate that positive horizontal spillovers at the regional level can be reinforced through agglomeration effects.

To sum up, regional effects in transition economies exist potentially due to vertical or horizontal effects as well as agglomeration effects. They exist independent from external effects at the national level. Given the small number of studies, this conclusion should be treated cautiously. However, this type of technology spillover through foreign subsidiaries is a potentially promising avenue for future research, in particular in relation to industrial clusters.

Summing up the finding from the meta-analysis

Figure 3.1 summarises the findings from our sample of studies with respect to the relevance of different technology transfer types via foreign subsidiaries in the transition economies. Clearly the positive direct effects dominate, followed by positive backward linkage effects.

Forward linkage effects tend to generate negative effects. Intra-industry effects are mostly found to be not significant, although in some instances horizontal effects tested positive or negative. Effects at the regional level exist embracing horizontal and vertical as well as agglomeration effects. The pattern of technology transfer in the transition countries mirrors the findings from other regions. It is fair to say that the more advanced transition countries such as the Czech Republic, Hungary, Poland, and Slovenia tend to have strong evidence for positive direct effects, which is likely to be coupled with a positive potential for backward linkages. This seems not to be the case for more backward countries such as Bulgaria and Romania.

Table 3.3 indicates which determinants for internal and external effects have been tested in our sample of econometric studies for the transition economies. The studies focused mainly on absorptive capacity,

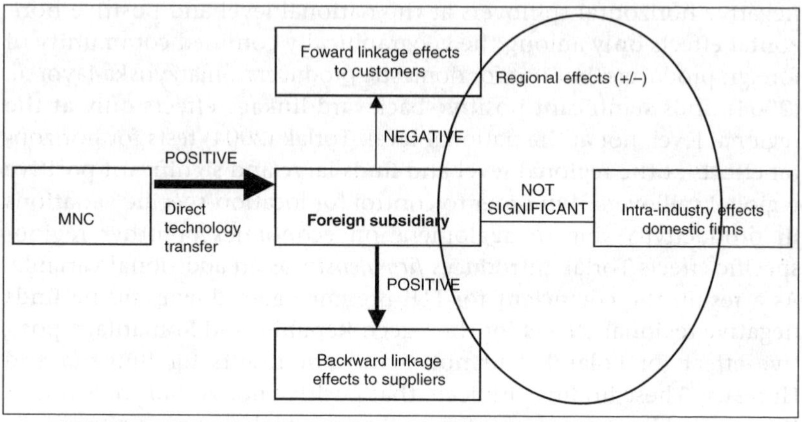

Figure 3.1 Results for technology transfer via foreign subsidiaries in the transition Economies

Note: The thickness of arrows/circle indicates the importance of the respective technology transfer types.

privatisation method, ownership characteristics and the competitive environment, as well as the motive and market orientation of FDI.

A number of conditions, including the impact of 'technology gaps' across countries/sectors and competition issues, have not been adequately discussed. The econometric approach towards technology spillover types can be enhanced by incorporating proxies not yet covered. Most work remains to be done in the field of direct and vertical effects. It emerges that, in particular, MNC-specific and subsidiary-specific determinants have not been taken into consideration. This seems to be a severe deficiency given the fact that these categories of determinants are particularly relevant for direct effects and backward linkages, which have been found the most important types of technology transfer in our sample.

It seems that in international as well as transition-specific research too much emphasis has been placed on horizontal effects. Existing research has very much concentrated on the question as to whether internal or horizontal effects exist or not, the determinants of the effects played a minor role. More recently, and particularly for the transition countries, vertical and regional effects have come onto the research agenda. To a considerable extent the findings for technology transfer via foreign subsidiaries are a result of the methodology applied. Direct transfer can be captured quite well by the production

Table 3.3 Determinants tested in econometric studies in the sample

Determinants	Direct/Internal technology transfer	Intra-industry spillover	Inter-industry effects: Forward	Inter-industry effects: Backward
Host country specific				
Technology gap,		X		
Factor endowment	X			
Property rights				
Policy factors (trade, local content, JV, SME, R&D, FDI, privatisation)	X	X		
National innovation system				
Exchange rate movemets				X
Industry-specific				
Type of sector (primary etc.)	X	X		
Type of industry	X			
FDI penetration/ MNC presence				
Speed of FDI penetration				
Absortive capacity in domestic firms		X	X	X
Technology intensity of sector				
Size and ownership characteristics of domestic firms		X		X
Competitive environment		X		
Capital and scale requirements				
Cluster formation/FDI penetration				
MNC-specific				
Motive for FDI	X	X		
Approach to subsidiary				
MNC market orientation (resource, efficiency, market, export)	X	X		X
MNC mode of organisation, international production networks				
Strategy in respect to R&D				

Table 3.3 Determinants tested in econometric studies in the sample (continued)

Determinants	Direct/Internal technology transfer	Intra-industry spillover	Inter-industry effects Forward	Backward
Technology transfer (adaption and application or innovation)				
Market position of MNC (e.g price sensitivity)				
Technology characteristics (age, complexity, tacitness, codifiability)				
Non-equity partnerships with local firms and motives				
Cost and risk of sourcing, proximity				
Subsidiary-specific				
Objectives and mandate				
Type of product (e.g. product cycle, standardisation)				
Role of subsidiary (autonomy, internal external initiative, functions)				
Size, age and ownership				
Mode of establishment (M&A, greenfield)	X			X
Type of R&D (applied, basic)				
Absorptive capacity		X		

Note: The X indicates that the determinant has been tested in empirical studies.

function approach having controlled for selection bias and endogeneity issues. The production function approach, however, seems to be incapable of modelling horizontal effects properly and alternatives such as non-linear estimation techniques or a combination with exit/survival rate estimations should be applied. The limited evidence for vertical linkages is also a result of slowly evolving methodology and insufficient data.

Moreover, the traditional production function approach in combination with quantitative data does not deliver any information about the actual *transmission mechanisms* of technology transfer at work. The insights into explicit technology transfer channels for the respective types of technology transfer are not yet properly understood. Technology transfer is treated as a 'black box' phenomenon. Taking MNC- and subsidiary-specific determinants into consideration could potentially deliver insights about the role of strategy and management within the MNC network in relation to direct and vertical technology transfer. These kinds of determinants, originating from the international business literature, are much closer to the actual technology transfer process. Therefore, the next section introduces evidence from the mezzo-analysis[15] to enhance our understanding of technology transfer in the transition economies.

Mezzo-analysis in technology transfer research

The methodology of mezzo-analysis

This research methodology combines qualitative information with quantitative analysis. In that respect, it stands between the standard econometric analysis and qualitative methods such as case studies or deep-level interviews. Mezzo-analysis builds on the results from quantitative work in the field of technology transfer. However, it allows us to integrate first-hand information with regard to the determinants and explicit channels of technology transfer types. Moreover, mezzo-analysis offers the opportunity to test international business concepts such as the developmental subsidiary and their relevance for technology transfer. In respect to the incorporation of qualitative data, there emerges a problem with the subjectivity and generalisation of evidence. However, when data is based on a large-scale survey, the generalisations should be more appropriate in comparison to cases studies or deep-level interviews. If the qualitative data are representative, they can be combined with official statistical data.

This approach has only recently been applied in the field of technology transfer. All the papers below discussed originate from authors organised in the framework of a European Union Research project.[16] The research is based on a survey from 433 manufacturing FIEs from Estonia, Hungary, Poland, Slovakia and Slovenia. The sample is dominated by Polish firms and there exist significant differences in sectoral distribution and representativeness per country. Most of the sample FIEs were established in 1993, are majority- foreign-owned, and by and large are intermediate rather than final producers (Majcen, Radošević and Rojec 2003b: 6–7).

Results from the transition countries

Majcen Radošević (2003b) suggest a conceptual model where the magnitude of productivity growth depends on the upgrading of the position of subsidiaries in the MNC network.[17] The subsidiary can basically upgrade its position by expanding its mandate or by adding a new business function. The underlying assumption is that increased autonomy can potentially deliver higher productivity – i.e. direct technology transfer. Majcen, Radošević and Rojec (2003b) assess the determinants of FIE productivity growth for five CEEs in an ordered probit regression model. The independent variables include subsidiary autonomy indicators for three groups of business function – namely, marketing, product and accounting and finance.[18] The estimation results show that the subsidiaries' productivity is positive and significantly determined by the size of foreign equity share, the size of the subsidiary and the proportion of exports to the parent company or other foreign firms. Foreign subsidiaries belonging in high- technology-intensity sectors exhibit significantly lower productivity growth than subsidiaries in other sectors. It was found that productivity growth is negatively correlated with the level of the foreign parent's control of operational management, and positively with the level of foreign parent's control of strategic management. It follows that the more subsidiaries are integrated into the foreign parent companies – equity, strategic management, export – the higher the productivity growth (2003b: 21). Moreover, it can be argued that strategic management-related business functions serve as explicit channels of direct technology transfer for FIEs in the transition countries. These findings suggest that a headquarter-centred approach fosters faster productivity gains opposed to a developmental approach with higher autonomy for the subsidiary. Therefore, the conceptional framework developed by Majcen, Radošević and Rojec (2003c) must be questioned, since productivity improvements seem negatively correlated with upgrading in terms of scope. In international studies, it has been found that granting strategic independence to subsidiaries may reduce the ability of headquarters to control their resources and thereby reduce the efficiency of the internal capital market and may reverse the localisation benefits of subsidiaries (Mudambi, 1999, as cited in Paterson and Brock (2001: 150). Similarly Egelhoff, Gorman and McCormick (1998) suggest that strong headquarters assignments are better than weak ones while at the same time suggesting that initiative should be encouraged. Certainly the research shows that a number of MNC- and subsidiary-specific determinants are relevant for direct technology transfer.

Majcen, Radošević and Rojec (2003a) find that Slovenian subsidiaries are more integrated into the foreign parent company's network via sales than via supplies while, on the other hand, they are more integrated into the Slovenian economy via supplies than via sales. This indicates higher potentials for backward linkage effects, but less potential for forward linkage effects, which would be in accordance with the general findings for vertical linkage effects from the meta-analysis. The degree of trade integration with the MNC and export propensity are likely to be determinants for inter-industry technology transfer. More research is necessary regarding which kind of supplies are sourced locally and which abroad, because it could be hypothesised that the higher the complexity of locally supplied goods or services, the higher the potential scope for technology transfer.

Summing up the findings of the mezzo-analysis

Bearing these methodological caveats in mind, the mezzo-analysis has shown that direct technology transfer in CEE foreign subsidiaries is positively associated with foreign parent control in the strategic management function and subsidiary autonomy in operational functions. In general, MNCs operating in CEECs seem to prefer a headquarter-centred approach to foster the growth of their affiliates. Strategic management-related business functions serve as explicit channels of direct technology transfer via FDI. Moreover, direct technology transfer is positively affected by the size of the foreign equity share, the size of the subsidiary and the export propensity to the parent company or other foreign firms.

It can be argued that the mezzo-analysis has made a considerable contribution in terms of testing the relevance of MNC- and subsidiary-specific determinants of direct technology transfer (see Table 3.4). Yet a variety of such determinants needs further consideration. This includes MNC-specific factors such as the strategy towards R&D, the characteristics of technology transferred (age, complexity, etc.), and the cost and risk of local sourcing, etc.

Moreover, subsidiary-specific factors such as the type of technology transferred (basic or applied), the type of product or (importantly) the absorptive capacity should be subject to further research. Importantly, the relevance of such determinants for external effects remains unexplored. The next step in the research could be to compare the pattern

Table 3.4 Determinants of technology transfer (meta- and mezzo-analysis)

Determinants	Direct/Internal technology transfer	Intra-industry spillover	Inter-industry effects	
			Forward	**Backward**
Host country specific				
Technology gap		X		
Factor endowment	X			
Property Rights				
Policy factors (trade, local content, JV, SME, R&D, FDI, privatisation method)	X	X		
Exchange rate movements				X
National innovation system				
Industry specific				
Type of sector (primary, etc.)	X	X		
Type of industry	X			
FDI penetration/MNC presence				
Speed of FDI penetration				
Absortive capacity in domestic firms		X	X	X
Technology intensity of sector	X			
Size and ownership characteristics of domestic firms		X		X
Competitive environment		X		
Capital and scale requirements				
Cluster formation/FDI penetration				
MNC-specific				
Motive for FDI	X	X		
Approach to subsidiary	X			
MNC market orientation (resource, efficiency, market, export)	X	X		X
MNC mode of organisation, international production networks	X			
Strategy in respect to R&D				

Table 3.4 Determinants of technology transfer (meta- and mezzo-analysis) (continued)

Determinants	Direct/Internal technology transfer	Intra-industry spillover	Inter-industry effects	
			Forward	Backward
Technology transfer (adaption and application or innovation)				
Market position of MNC (e.g price sensitivity)				
Technology characteristics (age, complexity, tacitness, codifiability)				
Non equity partnerships with local firms and motives				
Cost and risk of sourcing, proximity				
Subsidiary-specific				
Objectives and mandate	X			
Type of product (e.g. product cycle, standardisation)				
Role of subsidiary (autonomy, internal external initiative, functions)	X			
Size, age and ownership	X			
Mode of establishment (M&A, greenfield)	X			X
Type of R&D (applied, basic)				
Absorptive capacity				

Note: The X indicates the contributions of the *meta*-analysis. The **bold X** indicate the contributions of the *mezzo*-analysis.

of FIE–parent relationships and the resulting technology transfer effects in the transition countries with subsidiaries in the developed and the developing countries.

Notes

1. If MNCs capture market share, then local firms may under-utilise existing capacity in the short run. Although local firms will eventually redeploy slack resources, production function estimation will interpret non-utilised resources as a productivity loss in the short run
2. E.g. $Y_{ij}^t = A_{ij}^t F(K_{ij}^t, L_{ij}^t, M_{ij}^t)$, where $_i$ denotes the domestic firm, $_j$ the industry, and ^{t} the year. It is assumed that the production function is homogeneous of degree g in inputs and that it is both increasing and concave in all its arguments. The term A_{ij}^t measures TFP or the SR, which is assumed to vary across both firms and sectors and over time (as in Sinani and Meyer, 2004).
3. In the Heckman procedure, the selection issue is dealt with as an ordinary specification bias arising due to the omitted variables problem. Heckman proposes using estimated values of the omitted variables (which when omitted from the model give rise to the specification error) as regressors in the basic model (Damijan *et al.*, 2002: 195).
4. Here, unobserved productivity can be identified from the firms' observable variable input choice (e.g. material inputs), which is modelled as monotonic function of the capital stock and the unobserved (to the econometrician) productivity shock. A positive side effect of the Olley and Pakes (1996) technique is that it also facilitates a correction to offset the potential bias in the panel introduced by the non-random exit of firms due to bankruptcy, merger, or firms simply choosing not to report. Evenett and Voicu (2001) argue that such a bias or sample attrition effect is otherwise often sidestepped by the creation of a balanced panel.
5. The paper by Evenett and Voicu (2002) is also interesting from the point of view of a methodology for measuring direct effects. They demonstrated that the results for direct effects differ enormously depending on whether or not selection bias, non-random exits and sector effects have been properly controlled for.
6. As outlined above studies such as Smarzynska (2002) include a Herfindahl index (HI) to control for industry concentration, but they do not measure directly the competition between foreign and domestic enterprises. The approach by Kosova (2003) to test for the impact of foreign entry on the exit of domestic firms has also been employed by Görg and Strobl (2001) as well as De Backer and Sleuwaegen (2003).
7. Kosova suggests that one explanation for this finding is that the foreign growth rate represents export market creation or that the measurement picks up vertical inter-industry effects. External technology transfer could thus be subject to the extent to which the foreign subsidiary and the domestic firm compete in the same market.
8. If the share of foreign share in domestic output is 1 percentage point higher, the productivity growth factor for the local firm is 0.68% lower. Djankov and Hoekman (1998) find that each 10 percent increase in the

foreign asset share is associated with a 1.7 percent fall in the sales growth of domestic firms in the Czech Republic.

9. With low foreign penetration of the industry, the horizontal spillovers are low but increasing as foreign penetration increases. After some point foreign penetration of the industry may start dampening the activities of local firms, which cannot any long er compete with foreign affiliates and are forced to exit.
10. Measured in terms of the share of exports in output.
11. The division into sub-samples is also a potential way to deal with non-linearity in respect to horizontal spillovers, if the break in time is known. Alternatively, one could try to detect structural breaks in the data and run separate estimations, if the sample is wide enough. In that way perhaps the end of the shake-out period could be differentiated from the technology transfer phase.
12. Measured as the fraction of output produced by foreign firms in the total output of a particular region.
13. This finding is most likely due to too many missing observations for Poland in the sample.
14. Torlak (2004) does not control for selection bias nor endogeneity in his sample.
15. The concept of mezzo-analysis should not be confused with the term 'meso perspective' which refers to a level of economic aggregates between micro and macro.
16. The 5th Framework Programme Project: 'EU Integration and the Prospects for Catch-Up Development in CEECs'; WP4 'Mapping the Technology Structure of Branch Plants and Technology Integration of CEEC'.
17. The conceptual model is based on two forms of upgrading of the position of subsidiaries and on several dimensions of the integration of subsidiary into the MNC network. Upgrading of a subsidiary can occur through the introduction of new functions/mandate (e.g. sales, manufacturing, finance, etc.) as well as expansion of the existing functions. The subsidiary can also grow via an increase in scale (e.g. increase in sales, exports, extension of products) without changing the mandate. The upgrading of a subsidiary occurs through several dimensions (product flows, knowledge flow, equity changes). These dimensions can be analysed in terms of their intensity and direction between the headquarters and the subsidiary (Majcen, Radošević and Rojec 2003b: 4).
18. The three groups have been determined by factor analysis and contain (a) in marketing variable (marketing; market research; advertisement; distribution and sales; after-sales services; determining the product price); (b) a product variable (product development; process engineering); (c) an accounting and finance variable (accounting and financial operations; investment finance; supply and logistics).

4

Conclusions Drawn from the Latest Research: Lessons, Limits and New Research Trends

Björn Jindra

We have reviewed theory and evidence about the relationship between inward FDI and technology transfer in the transition countries. The focus was on technology transfer as a microeconomic phenomenon so the research concentrated on direct technology transfer from the MNC to the foreign subsidiary and external effects from the foreign subsidiary to domestic firms in the host economy. In our analysis we distinguished between direct, horizontal, vertical (backward, forward), and regional effects and outlined a number of explicit technology transfer channels via inward FDI taken from the theoretical literature. We proposed a new classification of determinants into host country-, industry-, MNC- and subsidiary-specific factors, according to the type of technology transfer.

Lessons from existing research

The theoretical framework was contrasted with the empirical evidence. A simple form of meta-analysis was used to interpret nineteen econometric studies, which mainly used the production function approach and firm-level panel data from the transition countries. The results clearly show that direct technology transfer is the most important channel for technology transfer to the transition countries. It is followed by positive backward linkages which on average prove to be larger than positive horizontal effects, where they exist at all. However, in most cases the horizontal effects tested not significant. Forward linkages were found to generate negative effects. Regional effects existed independent from indirect effects at the national level. However, the results for vertical and regional effects have to be treated carefully due to the limited number of econometric tests in respect of these technology transfer types in our sample. The general coverage of technology transfer determinants focused on absorptive capacity, privatisation method, ownership characteristics, the competitive environment, as well as the motive and market orientation of FDI. The pattern of technology transfer in the transition countries mirrors the findings from other international studies. It is probably fair to say that the more advanced transition countries, which are also in geographicae proximity to the Western European market – such as Czech Republic, Hungary, Poland the as well as Slovenia tend to have consistently strong evidence for positive direct effects, which is likely to be coupled with positive backward linkage effects. This seems not to be the casc for the less advanced countries in the sample, such as Bulgaria and Romania. The main contribution of evidence from the mezzo-analysis is the finding that the more CEE foreign subsidiaries are integrated into the MNC network, the higher is the likelihood of direct technology transfer. Moreover, mezzo-analysis found strategic management-related business functions as an explicit channel for knowledge transfer and proved that direct technology transfer is affected by subsidiary-specific factors such as foreign equity share, size of the subsidiary, and export propensity to the parent firm.

Limits of existing research and new research trends

Traditional econometric studies did not succeed in differentiating explicit channels for types of technology transfer. Here, evidence from mezzo-analysis was able to give some initial insights. The coverage of empirically tested determinants for technology transfer lags far behind the number of theoretically developed determinants. A number of factors are inadequately discussed, including the impact of 'technology gaps' and competition, at both national and industry level.

The results obtained are to a considerable extent related to the limitations of the methodological approach taken and the econometric problems involved. Different data and estimation techniques influence the results. Despite the fact that the traditional econometric studies focused heavily on horizontal effects, the production function approach seems not capable of modelling the dynamics of technology transfer appropriately. In order to differentiate the crowding-out from the horizontal technology transfer effect, the combination of the standard approach with a model to estimate exit/survival rates of domestic firms is proposed. This could potentially explain the general absence of horizontal effects.

The results from our study suggest extending existing research into the area of direct and vertical linkage effects, which prove to be most relevant for technology transfer to the host economy. In particular, the explicit channels as well as MNC- and subsidiary-specific determinants of the respective technology transfer types should be subject to further research. Research on regional effects and agglomeration economies in the context of industrial cluster could also be. The traditional production function approach has to be further complemented with mezzo-analysis and qualitative evidence; otherwise it will not be possible to generate a better understanding of the nexus between the changing nature of MNCs and the resulting technology effects for the host economy.

The broader research agenda

The general dominance of internal effects over external effects raises the question of how much of the overall effects is actually appropriated by domestic consumers and producers or the MNC. In other words, who actually benefits from FDI? An answer to this question requires us to take into consideration not only productivity impacts but also changes in competition and prices. Following from this, one could ask whether FDI subsidies and other FDI-attracting incentives are justified or not. An important aspect which could not be considered in this chapter is the question as to whether there is productivity convergence or divergence between the MNC and the foreign subsidiary as well as between the foreign subsidiaries and domestic enterprises. The worst-case scenario would be that there is productivity convergence in the former but not in the latter case, giving rise to a dual structure in the host economy. Leading from this, one could contribute to the research on the links and causality between inward FDI and economic growth. Setting these results in the context of the determinants of internal and external effects could permit us to draw conclusions about the relevance of foreign subsidiaries and MNC strategies, as well as host country factors, in explaining differences in the catch-up growth process across the transition countries.

Part II
Empirical Studies

Part II of the book contributes two innovative empirical studies on the role of FDI for technology transfer in CEECs, and is preceded by a brief overview of the role of FDI and its effects in the CEE economies of. The brief introduction is explicitly comparative across a selection of five countries in the region – Estonia, Hungary, Poland, the Slovak Republic Hungary and Slovenia; some data are available for only a sub-set of countries.

5

FDI, Productivity and Economic Restructuring in Central and Eastern Europe

Judit Hamar and Johannes Stephan

This introductory chapter to Part II of the book presents a comparative overview of economic development, and the changing conditions for and results of FDI as a mechanism of productivity growth in Estonia, Hungary, Poland, the Slovakia-Republic, Slovenia. By summarising briefly the main similarities and differences between these countries (with a particular focus on Hungary), we try to determine whether differences by countries depend on their different stages in FDI attractiveness, labour productivity,[1] economic development levels and restructuring by technology intensity.

International theory of trade and finance

Our first hypothesis relates to the international theory of trade and finance and the empirical evidences on the effects of FDI on the host country:

- FDI speeds up productivity growth and restructuring.
- FDI effects differ strongly across countrie, due to:
  - Differences in the initial economic and political conditions.
  - Differences in the timing and sequencing of policy changes (setting of legal conditions, liberalisation of foreign trade and FDI, methods and timing of privatisation, etc).
  - Differences in factor endowments (past specialisation and capacity for adjustment).
  - Differences in the prospects for future catch-up, as perceived by international markets. Perceptions are determined by the different

productivity levels the countries have reached, by the extent and speed of technological upgrading and by the different levels of penetration of the markets by existing FDI.

Our research results point in the same direction: the size of effects of FDI on the host economies depends on *country-specific* features. This limits the comparative nature of our analysis. Since the outset of systemic transformation in the early 1990s, the countries assessed in our project were able to narrow their productivity gaps to levels predominant in Western Europe. Still, large differences exist, both *vis-à-vis* the West and also between our CEE countries. The inflow of FDI played an important role in each of our countries in their individual catching up process, but the effects differed in the time of entry, in the activities carried out and by the timing and sequencing of policy reforms.

Figure 5.1 demonstrates that the growing stock of FDI (here normalised by the host country's GDP, for 1993, 2000 and 2002) is positively associated with productivity improvement in each of the countries. However, the slopes of the curves tend to become flatter the further right the curve is located on the Figure 5.1.

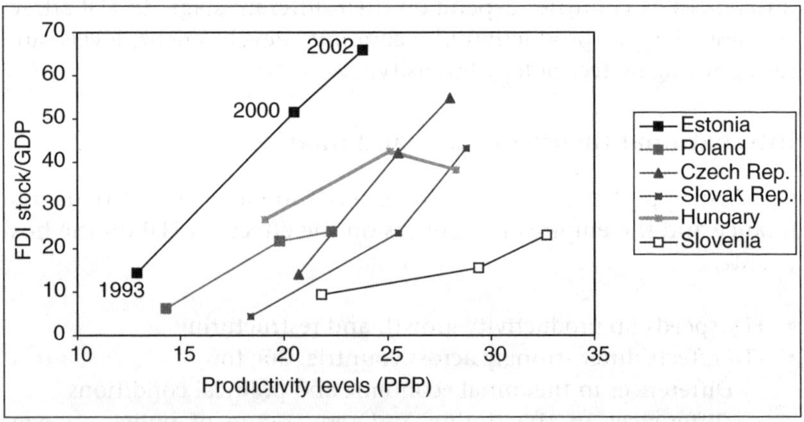

Figure 5.1 FDI and national labour productivity, 1993, 2000 and 2002

Notes: Labour productivity levels calculated with PPP exchange rates in EUR 1,000 value added per employment (full-time equivalent); FDI stocks are calculated as inward FDI (net of outflows) stock as a percentage of GDP.
Sources: Productivities: Vienna Institute for International Economic Studies (WIIW), National Statistical Offices, own calculations. FDI stock: UNCTAD, FDI/TNC database (World Investment Report, 2003).

- The highest increases in terms of both productivity and in FDI occurred in Estonia: from the comparatively lowest productivity level with a relatively high FDI ratio, productivity growth was the strongest and, comparably, additional FDI inflows increased the most, indicating the greatest FDI attractiveness.
- At the other extreme, Slovenia started with a high national productivity level, but while productivity growth was strong, the country's attractiveness for additional FDI was low, and little further FDI inflows were recorded.
- the Slovak and Czech Republics started with productivity levels somewhere between these two extremes, and their additional increases in FDI inflows were similarly between those of Estonia and Slovenia. In the Slovak Republic, increases in FDI inflows were initially lower which might be associated with political uncertainty.
- Hungary and Poland took a different path. Hungary has a much longer history of FDI than any of the other countries so, assuming decreasing returns, FDI attractiveness is lower than anticipated according to the criterion of initial productivity levels (the negative association between 2000 and 2002 is due to the repatriation of profits from FDI which were registered as negative FDI inflows, just another sign of comparatively mature FDI). Poland should be attractive becauseof a low initial productivity level, yet FDI inflows per GDP are rather low, possibly due to the country's sheer size.

National and industrial labour productivity levels

Labour productivity increases in CEE were partially due to falling employment rates in the second part of the 1990s. The only exception was Hungary where, in spite of a shrinking population and despite slightly increasing employment levels, the employment rate still remained the lowest among the countries assessed, even at the end of the 1990s.

Each of our countries narrowed the national productivity gap to the EU-15, but in the second half of the 1990s this development was partially due to a slowdown in labour productivity improvement in the EuropeanUnion itself (*vis-à-vis* the United States, the EU-15 productivity level also fell, from 79.5 per cent in 1995 to 73 per cent in 2001). By 2000, national labour productivities in each of our countries surpassed country with the lowest level in the EU-15, which was Portugal. All CEEC except Poland have already surpassed the of threshold 50 per

Table 5.1 National and industrial labour productivity levels, selected CEECs, 2000

% of average EU-15	Average national levels	Average industrial levels
Czech Rep.	62.7	51.3
Estonia	52.0	34.2
Hungary	61.1	51.6
Poland	46.3	44.5
Slovak Rep.	58.6	43.9
Slovenia	76.9	58.6

Note: PPP, €, sum of value added in all sectors (without taxes, subsidies and Financial Intermediation Services Indirectly Measured (FISIM)) per total employment (not accounting for intensity of use).
Sources: EUROSTAT, OECD, WIIW, National Statistical Offices, own calculations.

cent of the productivity gap to the EU-15, and the Slovenian productivity level has already surpassed that of Spain.

The CEECs' industrial productivity levels, however, demonstrate a much wider gap from the EU-15 average (see Table 5.1) than the gap in the national levels. This is mainly due to the fact that productivity levels in the financial sector in particular, and also in market-and household-related services, are much closer to the EU-15 average. In general, CEECs' industries today still exhibit a sizeable gap in industrial labour productivities, achieving no more than 35–60 per cent of the EU-15 average. Productivity growth by far outpaced that of the EU-15 average, but even those above-average rates will still be associated with several decades of productivity convergence. In estimations for the time needed for complete catch-up, the relevant literature assumes some 10–15 years for Slovenia, 20–30 years for the Czech Republic,Hungary and the Slovak Republic, and even longer for Estonia and Poland.

The differences between national and industrial productivity levels are associated with fast restructuring in all countries: the weight of agriculture in the CEE economies fell rapidly everywhere (except in Poland, by employment shares), while service sectors grew rapidly in all the CEE countries assessed in our project. The service sector had already become the largest one in Hungary in 1993 and had grown further to 60% in terms of employment and 63% in terms of value added by 2000. The manufacturing sector retained its weight in Hungary and increased in Estonia in terms of employment, while by gross value added Slovak Republic manufacturing increased its contribution to GDP the most (see Table 5.2).

Table 5.2 Structural adjustments: the share of manufacturing sectors, by employment and by gross value added,1993 and 2000[a]

% of GDP	Estonia		Poland		Czech Rep.		Slovak Rep.		Hungary		Slovenia	
	1993	2000	1993	2000	1993	1999	1993	2000	1993	2000	1993	2000
By employment	21.4	22.6	20.2	17.6	31.2	30.0	26.8	25.7	24.5	24.2	36.6	29.4
By value added	19.0	16.5	22.0	20.6	25.8	26.3	20.6	24.0	22.0	23.5	29.5	27.2

Note: PPP, €, sum of value added in all sectors (without taxes, subsidies and FISIM) per total employment (not accounting for intensity of use).
a Czech Republic data is 1999.
Sources: see Table 5.1.

In all the countries observed, the structural content of the productivity gap decreased, but it remained large enough to be a problem: these countries will need a long time to catch-up with the Western EU average. The gap between the levels of national labour productivities and the EU-15 average – and in particular that of industrial labour productivities – remains large in all cases. That means that further intense restructuring – and, in particular, productivity improvements – will be needed in the industries of each country to close the productivity gap to the West. In some countries – such as Estonia and Hungary the Slovak Republic efficiency improvements in the public sectors will also be inevitable.

For all the countries observed here, we can detect that FDI played a similarity important role for productivity upgrading, but the level of importance of FDI, the speed of restructuring, the economic structure by activities, the industrial structure by factor intensities and the respective productivity levels of countries and sectors remain highly differentiated even today.

FDI assisted rapid restructuring and economic development in all CEECs, yet the countries attracted foreign investors to a very different extent. FDI inflows also varied strongly over time (see Figure 5.2).

Inflows of FDI into the CEE are determined to a large extent by the privatisation of formally state-owned enterprises (SOEs), but foreign investment also depends on current perceptions about the attractive-

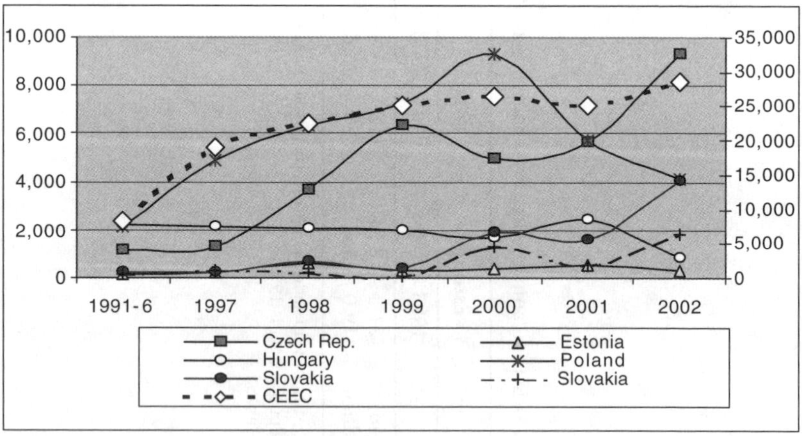

Figure 5.2 FDI inflows into CEECs, by host region, 1991–6, 1997–2002 million US$

Note: TNC = transnational corporation.

Table 5.3 Inward FDI stock as a percentage of GDP, 1995–2002

Country	1995	2000	2001	2002
Czech Rep.	14.1	42.1	47.4	54.8
Estonia	14.4	51.5	57.2	65.9
Hungary	26.7	42.5	45.4	38.2
Poland	6.2	21.7	22.4	23.9
Slovakia	4.4	423.6	30.4	43.2
Slovenia	9.4	15.5	16.4	23.1

Sources: UNCTAD, FDI/TNC database (*World Investment Report*, 2003).

ness of the host economy (extent of reforms, economic restructuring, political stability, etc.) and future expectations of investors about profitability. Hungary was the first CEE country to provide the conditions to attract large amounts of FDI, so net FDI inflows were robust from at least 1996 (even if they tell off in 2002 due to increasing profits repatriation). Poland, as the largest economy attracted nearly as much FDI as Hungary between 1991 and 1996. Inflows then became much larger, yet fell dramatically from 2000 onwards. Due to a late start in privatisation by foreign investors and the country's financial crisis, FDI inflows in the Czech Republic picked up only in 1997 and grew rapidly until 2002. In the Slovak Republic and Slovenia, FDI inflows started to increase significantly after 1999, in parallel with a more FDI-friendly policy.

FDI stock as a percentage of GDP reached (Table 5.3) its highest rate in Estonia after 2000 and then in the Czech Republic after 2001. Even the Slovak Republic had a higher rank in 2002 than Hungary which was the only country where the FDI/GDP ratio fell seriously in 2002.[2] The cumulated stock of FDI in Poland remained (after Slovenia) relatively the lowest.

The importance of FDI in the CEE economies

The CEECs are at different stages concerning the importance of FDI in their economies:[3]

- FIEs in Slovenia accounted for only 15 per cent of total assets and 12 per cent of employees of the Slovenian non-financial corporate sector in 2001. They realized 20 per cent of sales, 22 per cent of operating profits and 31 per cent of total exports.
- In Poland, the number of FIEs in the economy grew rapidly between 1993 and 2001 (from 15,814 to 44,477), employing 310,000 persons

Table 5.4 FIEs' share in the Hungarian economy, 1992–2001 (per cent)

Item	1992	1996	2000	2001
No. of FIEs	21.4	19.4	14.4	12.6
Assets	17.8	44.7	63.6	64.9
FDI/Assets	10.1	31.6	56.7	58.4
Investment	n.a.	61.8	60.1	59.1
Sales	24.4	47.2	54.1	53.3
Exports	37.3	70.8	82.5	83.1
Employment	15.3	28.7	32.3	30.5
Salaries	19.1	41.3	47.2	44.7

Sources: Tax reports of double-entry accounting firms, own calculations.

in 1993 966,000 in 2001 (8.4 per cent of total employment, except for agriculture). FIEs' share in exports grew from 16 per cent to 53 per cent, and had reached 61 per cent in total imports by 2001. The output share of FIEs increased from 12.4 per cent in 1994 to 34 per cent in 2001, and in terms of investment from 3.3 per cent to 30.4 per cent.

- In Hungary, the number of FIEs grew from 12,363 in 1992 to 20,943 in 2000[4]. Whilse relative to all firms investment did not increase after 1996, in assets and especially in exports the role of FIEs grew steadily (see Table 5.4). By the end of the 1990s, FIEs dominated the economy. 2000, however, was the first year when indigenous firms invested relatively more than FIEs (the investment share of FIEs fell from 71 per cent in 1999 of the previous year to 60 per cent) and by 2001 FIEs' share had diminished almost in every respect (except for the share in total assets).
- In Estonia, FDI also played a very important role in economic restructuring and created a favourable base for real convergence toward the European union. The number of FIEs was 3,066 at the end of the 1990s.[5] The ratio of FDI to GFCF accelerated in the second part of the decade: from 24 per cent (as an annual average between 1991 and 1996) to 31 per cent (1997–2001). In 2002, however, FDI inflows relative to GFCF fell to 17 per cent.

The role of FDI with respect to GFCF (Table 5.5) accelerated in each country during the second part of the 1990s, except for Hungary, where it was highest in the first part of the decade.

Table 5.5 Inward FDI flows as a percentage of gross fixed capital formation (GFCF), 1991–2002

GFCF = 100	1991–6	1997–2001	2002
Czech Rep.	9.6	28.3	59.1
Estonia	23.9	30.5	16.8
Hungary	26.8	18.3	n/a
Poland	10.1	17.4	11.4
Slovakia	4.4	15.2	56.9
Slovenia	4.0	5.6	37.1
CEEC	**5.8**	**14.9**	**17.2**

Sources: UNCTAD, FDI/TNC database (*World Investment Report*, 2003).

The changing focus of foreign investors, by broad sectors of the economy

The distribution of FDI by main economic activitiy also differed by countries. Manufacturing and trade were the focus of foreign investors at the early stage of FDI inflows in Hungary and in Estonia (as now in Slovakia and Slovenia). The timing of large privatisation processes also influenced the structure of FDI inflows.

- In Hungary, the FDI ratio to assets by sectors had reached 21 per cent in manufacturing, and 15 per cent in trade in 1992, and increased to 51 per cent and 36 per cent, respectively, in 1996 (Table 5.6). Since then, its dynamics have slowed down: the FDI: assets ratio grew to only 58 per cent in manufacturing, and to 56 per cent in trade, while business services and especially financial activities had by 2000 become the most attractive sectors for FDI. The FDI: assets ratio grew from 21 per cent to 58 per cent in business services, and from 44 per cent to 89 per cent in financial intermediates between 1996 and 2000.
- The changing FDI distribution by sectors showed a similar trend in Estonia as in Hungary. During the first years of transition, the structure of inward FDI was rather stable. Manufacturing industry was the major recipient, followed by the wholesale and retail sales sectors. These two sectors of the economy attracted nearly 70 per cent of the FDI inflow in 1994–5. Since 1996, the importance of the manufacturing industry as a target for FDI decreased and the financial sector, together with transportation and communication industries, became more attractive to foreign investors. Changes in

Table 5.6 Stock of FDI in manufacturing and some major service sectors, 1998, 2000–2

Activities, sectors NACE	Estonia (2002)	Hungary (2000)	Slovak Rep. (2002)	Slovenia (2001)	Poland* (1998)
D Manufacturing	18.7	19.0	41.7	38.5	69.6
E Electricity, gas and water supply	2.5	5.0	0.2		
G Wholesale, retail trade	13.4	8.6	12.9	13.9	5.3
I Transport, storage, communication	22.6	4.2	13.2		
J Finance	27.4	36.6	26.5	27.8	17.6
K Real estate, renting and business activities	9.6	23.5	3.2	11.1	
Total	**100**	**100**	**100**	**100**	**100**

*Industry includes mining and electricity, trade includes real estate, services attracted 25% of FDI in 1998, but it included construction, too.
Sources: For Poland, the source is the Polish Agency for Foreign Investment; for Hungary, national tax data base and own calculations; for the other countries calculations are based on country studies (http://www.iwh-halle.de/projects/productivity-gap).

the structure of FDI inflows reflected the dynamics of the privatisation programme of the Estonian government. After 1996, there were no large privatisation projects in industry, while some infrastructure and transport enterprises attracted heavy investment, as did two leading commercial banks (involving Swedish and Finish investments in 1998 and 2002).

The changing focus of foreign investors by activities after a longer involvement and the differences in the weight of FDI in manufacturing can be illustrated by the distribution of FDI stocks accumulated in manufacturing industries (see Table 5.6).

The penetration of CEECs' manufacturing industries by FDI

The share of FIEs in manufacturing industry is also very different across countries:

- By the end of the 1990s, the highest (perhaps already extreme) dominance of FIEs was reached in Hungarian manufacturing industry,

Table 5.7 FIEs' share in manufacturing industries, 2000 and 2001

FIE in all manufacturing firms (%)	Estonia	Hungary	Slovenia	Poland
	2000		2001	
No. of firms	9.8	17.1	4.8	1.2
Fixed assets	37.2	75.6	21.8	13.4
Sales	34.3	73.7	26.2	21.1
Exports	44.9	88.5	33.8	50.9
Employment	27.3	49.3	16.5	21.0

Sources: See Table 5.6.

especially pronounced in terms of exports (Hungarian data in Table 5.7 relate to double accounting entry firms only), sales and capital endowment, but also in terms of employment in foreign subsidiaries.

- In Estonia, the extent of foreign penetration in manufacturing industry in 2000 is indicated by the 10 per cent share of FIEs in the total number of manufacturing firms, by a 37 per cent share in fixed assets, by a 27 per cent share in employment, by a 34 per cent share in sales and by a 45 per cent share in exports.
- In the Slovenian manufacturing sector, the number of FIEs in all firms to talled less than 5 per cent, but FIEs employed 17 per cent of the manufacturing labour force, had 22 per cent of the fixed assets produced 26 per cent of sales and exported 34 per cent of total manufacturing exports.
- For the Slovak manufacturing industry, we have data only for the share of FIEs in total fixed assets, which reached 22 per cent in 2000.

The role of FIEs in CEECs' productivity improvements

The growing inflow of foreign technology and knowledge via FDI can be assumed to have helped the productivity improvement in the manufacturing industries of each of our CEE economies. The clear differences in main economic indicators between FIE and domestic DE enterprise (DE) groups can serve as an indication for this (Table 5.8). The performance gap between the two groups of companies narrowed somewhat by the end of the 1990s, but remained large, proving the advantageous position of the FIE group almost in all of the countries observed. The largest performance differences, however, still existed in

Table 5.8 Performance differences between FIEs and DEs, 2000 and 2001

Ratios in FIEs/DEs	Hungary	Estonia	Slovenia
	2000		2001
Sales per employee	2.9	1.26	
Value added per/employee	4.0	1.45	1.2
Wages	1.6	1.28	
Capital: labour	3.2	1.65	1.5
Exports: sales	2.8	2.17	1.4

Sources: See Table 5.6.

Hungarian manufacturing, while the smallest differences could be observed in Slovenia:

- In Estonia, convergence between FIEs and DEs could be registered in unit labour cost (with the ratio falling from 0.69 to 0.92 between 1996 and 2000), and in capital: labour ratios (from 3.39 to 1.65), while differences in export orientation remained high (in FIEs 2.33 times higher in 1996, and 2.17 times higher in 2000). Yet the difference in productivity between the FIE and DE groups by value added per employee shows an opposite trend: in 1996, it was 1.41 times higher in favour of FIEs, and in 1.45 times 2000, indicating that FIEs' productivity levels grew slightly faster than those of the DE group average. The ratio of wage levels also increased a little (FIEs paid 1.27 times higher wages in 1996, than DEs, and 1.28 times higher in 2000).
- In Hungary, similar to the Estonian trend, capital: labour ratios decreased from 3.9 in 1996 to 3.2 in 2000. While the difference in wage levels was higher than in Estonia, it converged in Hungary somewhat from (1.8 to 1.6) during the same period. The ratio of exports per employees was also higher and increased further (from 5.5 to 7.9). The indigenous firm group increased its profitability more than the FIE group between 1996 and 2000 (growth of profit after taxation in the DE group was 5-fold, while in the FIE group it was only 2.4-fold). Productivity differences between the two groups by net sales per employee were higher and increasing (2.6 times in 1996 and 2.9 times in 2000), while the ratio of value added per employee decreased from 5.9 times to 4 times.
- The differences between the FIE and DE groups were lowest in Slovenia: the ratio of value added productivity in manufacturing was only 1.2 in 2001. The ratio of assets per employees was 1.5,

while the difference in export orientation (export per sales) was only 1.44 – FIEs exported 72 per cent and DEs only 50 per cent of their sales. Compared to the Hungarian rates (where FIEs exported 60 per cent of their output and domestic firms only 22 per cent in 2000), it is clear that both FIEs and DEs were much more export oriented in Slovenian manufacturing industry.

- In Poland, overall productivity (measured in revenue per employee) in the economy grew faster than in the FIE group (annual average growth rate between 1996 and 2000 was 124.5 per cent in total and 123.1 per cent in the FIE group). The ratio of productivity differences between FIEs and DEs increased until 1996, but since then domestic firms have narrowed the gap: the FIE/DE ratio decreased from 2 times to 1.5 times. The export orientation of FIEs was also 2.6 times higher than that of DEs.
- For Slovakia, we can use only estimations. The productivity gap (by employment) was estimated to be 2.9 times lower in the FIE group as compared to the whole economy; by value added per employee, the rate was only 1.6 times. According to the profitability indicator of value added to sales, FIEs had lower than average results (0.8). FIEs' share in exports decreased from 37 per cent in 1994 to 31 per cent in 2000, and increased only in 2001 (to 36 per cent).

The role of FIEs in technology upgrading

Productivity differences and economic development in CEECs depended significautly on structural changes within their manufacturing industries: the prospect of catching-up was to some degree determined by a shift from declining activities (firms) to dynamic ones, and from low value added activities to high value added ones. The analysis of the path of specialisation and the changing industrial structure (e.g. by technology intensity) revealed that each of our CEECs attracted FDI first in labour-intensive low-technology production. A shift towards more sophisticated activities could later be registered in all CEECs, and was led mainly by FIEs, while DEs remained more traditionally specialised. In spite of this progress, however, even in the most advanced countries the FIE group was still mainly located in the less (low- and medium- low) technology- intensive industries. This is particularly true in terms of the number of firms and share of employees (see Tables 5.9 and 5.10).

The OECD classification of manufacturing industries includes four groups. High-*technology industries* include: Aircraft and spacecraft

Table 5.9 Structure of FIEs according to technology levels of manufacturing industries (WIIW classification)

Sectors (by WIIW classification)	Estonia	Hungary	Poland	Slovakia	Slovenia
	Distribution of no. of FIEs in manufacturing				
'High'- and medium-high'-technology	13.9	28.4	27.6	30.2	38.4
'Low'- and 'medium-low'-technology	86.1	71.6	72.5	69.8	61.6
Total (no. of FIEs)	**100 (402)**	**100 (3743)**	**100 (4417)**	**100 (258)**	**100 (302)**
	Distribution of FIEs, by fixed assets				
'High'- and 'medium-high' technology	12.1	49.5	37.7	21.4	47.8
'Low'- and 'medium-low' technology	87.8	50.4	62.3	78.6	52.2
Total (fixed assets of FIEs)	**100**	**100**	**100**	**100**	**100**
	Distribution of FIEs, by sales				
'High'- and 'medium-high' technology	18.5	55.3	34.3		59.0
'Low'- and 'medium-low' technology	81.5	44.8	65.6		41.0
Total (sales of FIEs)	**100**	**100**	**100**		**100**
	Distribution of FIEs, by exports				
'High'- and 'medium-high' technology	24.4	74.3	53.3		64.8
'Low'- and 'medium-low' technology	75.6	25.7	46.7		36.1
Total (exports of FIEs)	**100**	**100**	**100**		**100**
	Distribution of FIEs, by employment				
'High'- and 'medium-high' technology	25.4	55.3	42.6		48.5
'Low'- and 'Medium-low' technology	74.5	54.6	63.4		51.4
Total (employment of FIEs)		**100**	**100**	**100**	**100**

Note: The classification is based on the use of technology (WIIW classification).
Sources: For Hungary and Slovenia, tax reports of double-entry accounting firms. For Slovak Republic, estimation based on the sample firms, except the distribution of fixed assets of FIEs: here the source was the Statistical Office (own calculation based on the country reports data).

Table 5.10 Importance of FIEs, by technological level (OECD and WIIW classification)

Manufacturing	**Shares of all FIES of firms in the sector group (%)**				
Hungary, 2000 (NACE Rev. 1 OECD)	**No. of firms**	**Fixed assets**	**Sales**	**Exports**	**Employment**
'High'-technology industries	18.4	81.1	89. 9	96.8	69.7
'Medium-high'-technology industries	20.8	86.6	83.7	93.1	60.9
'Medium-low'-technology industries	18.0	77.6	70.3	78.1	47.1
'Low'-technology industries	15.2	58.5	56.2	71.6	38.6
Total manufacturing industries	**17.1**	**75.6**	**73.7**	**88.5**	**49.3**
Slovenia, 2001 (NACE Rev. 1 (OECD)	**No. of firms**	**Fixed assets**	**Sales**	**Exports**	**Employment**
'High'-technology industries		22.5	21.4	23.7	19.1
'Medium-high'-technology industries		31.5	44.2	49.1	25.3
'Medium-low'-technology industries		20.2	23.5	30.4	18.0
'Low'-technology industries		16.1	14.4	19.1	9.9
Total manufacturing industries	**4.8**	**21.8**	**26.2**	**33.8**	**16.5**
Estonia, 2000 (WIIW)	**No. of firms**	**Fixed assets**	**Sales**	**Exports**	**Employment**
'High'-technology industries	13.9	33.8	43.0	56.1	41.3
'Medium-high'-technology industries					
'Medium-low'-technology industries	10.6	54.1	35.7	44.2	22.6
'Low'-technology industries	8.6	32.5	31.8	41.4	24.9
Total manufacturing industries	**9.8**	**37.2**	**34.3**	**44.9**	**27.3**

Table 5.10 Importance of FIEs, by technological level (OECD and WIIW classification) (continued)

Manufacturing	**Shares of all FIES of firms in the sector group (%)**				
Poland, 2001 (WIIW)	**No. of firms**	**Fixed assets**	**Sales**	**Exports**	**Employment**
'High'-technology industries	1.9	14.8	25.8	59.8	30.0
'Medium-high'-technology industries					
'Medium-low'-technology industries	1.5	9.5	13.1	37.4	19.2
'Low'-technology industries	0.9	16.6	22.9	47.7	17.3
Total manufacturing industries	**1.2**	**13.4**	**21.1**	**50.9**	**21.0**

Notes: See Table 5.9.
Sources: See Table 5.9.

(3530); Pharmaceuticals (2423; 244); Office, accounting and computing machinery (30); Radio, television and communication equipment (32); medical, precision and optical instruments (33). *Medium-high technology industries include*: Electrical machinery and apparatus (31); Motor vehicles, trailers and semi-trailers (34); Chemicals excluding pharmaceuticals (24 excl. 2423); Railroad equipment and transport equipment (352 + 359; 35.2 + 35.4); Machinery and equipment (29). *Medium-low technology industries include*: Coke, refined petroleum products and nuclear fuel (23); Rubber and plastic products (25); Other non-metallic mineral products (26); Basic metals (27); Fabricated metal products, except machinery and equipment (28); Building and repairing of ships and boats (351; 351). *Low-technology industries include*: Food products, beverages and tobacco (15 + 16); textiles, textile products, leather and footwear (17 + 18 + 19); Wood, pulp, paper, paper products, printing and publishing (20 + 21 + 22); Manufacturing not else classified, and recycling (36 + 37).

The WIIW classification does not separate 'high-' and the 'medium-high' technology industries (the Aircraft and spacecraft, Pharmaceuticals, Building and repairing of ships and boats and the different Engineering sub-sectors).

Table 5.9 presents the shares of FIEs in sectors by four groups of technology intensity (OECD classification, where data is available, otherwise WIIW classification). The shares of FIEs in each sector demonstrate clearly how important a role FDI played in upgrading CEECs' technological structure: the more sophisticated the activities, the higher is the share of FIEs, especially in exports, sales and capital endowment. This is most explicit for Hungary, where FDI arrived earliest.

For the Slovak Republic, the project's country study reports only statistical office data for the FDI share of total investment in industrial production, but on the basis of estimations by the author, 'Despite the fact that FDI influenced the value added improvement and technological upgrading, value added per employee was only slightly more positive in the FIEs group than the economic average. The low ratio of value added of sales, and thus also low economic efficiency demonstrate that the main orientation of foreign investors in an early stage of development is focused on the most important comparative advantage of Slovakia – (a) cheap, technically educated and skilled labour force. This tendency is continuing, and only in some branches, sophistication of production is mildly increasing' Country study for the Slovak Republic (http://www.iwh-halle.de/project/productivity-gap)

Summing up the statements of the project's country studies about the motives of foreign investors, and the host country's move to attract them, more similarities than differences can be found. Each of the countries assessed had (more or less) the same comparative advantages at the beginning of FDI inflow liberalisation: a relatively cheap but well-educated labour force, knowledge of companies (after decades of outward processing trade, OPT) and geographical proximity to the main investors. Differences can be found in market size (in particular, Poland), in the rank of the main investors by geographical origin (Scandinavians in Estonia; French in Poland; German, Dutch and Austrians in Slovenia, Slovakia and Hungary; American and Japanese investors also displayed a higher activity in Hungary) and also in timing and sequencing of legislation and economic restructuring and stabilisation.

The main motives for investment in each country were at first market seeking (home and neighbouring markets, prospect of EU integration). Next, foreign investors were increasingly efficiency seeking (mainly relatively low labour costs, and in Poland and Slovakia, raw materials as well), combined with MNC strategies (global and/or regional). Only the Slovenian study mentioned the importance of recognized trademarks (however, this was an unspoken motive for investment in the food industry almost in all our countries).

The volume of FDI across all our countries depended significautly on the timing of *legislation* (the most important laws for creating safe legal conditions for FDI, such as the Foreign Trade Act and the Company Law in Hungary in 1988–9, or in Slovenia in 1999), and on the *sequencing* of privatisation projects. Several barriers to FDI were abolished in Slovakia in the early 2000s, while in Hungary, Estonia, and in Poland, it happened during the early 1990s.

The most important basic conditions everywhere, however, are the *political and economic stabilisation* (transparency and prospects, as the peaceful and fast transition in Hungary promised in the early 1990s, and economic stabilisation in 1995 facilitated; or the more recent consolidation of the political and economic situation in Slovakia). Clear regulations and special incentives to investors could attract more (and a lack of them would deter) FDI in competition with conditions in neighbouring countries. Any disturbance in political and economic transparency seems to have been an important factor in the recent decline in attracting FDI in Hungary, and is clearly indicated in the case of the Slovak Republic.

Differences in regulations (and in possible future incentives), however, will be eliminated by full EU membership, and the comparative advantages of relatively low labour costs are rapidly eroding in all of these economies (especially in comparison to the non-member neighbouring European or transition economies and most of the developing countries, such as China). Each of the candidate countries however has some need to catch-up with hourly average EU labour costs, even if we consider the levels of productivity gap. The relative level of hourly labour costs compared to the average of the EU-15 in

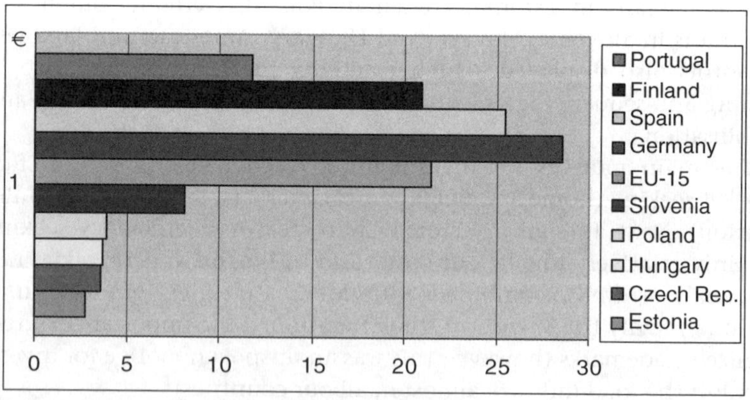

Figure 5.3 Hourly industrial labour costs in manufacturing industry, European countries, 2000

2000 was still 38 per cent even in the most developed of our countries, Slovenia; the Czech Hungarian and Polish levels were 17 – 18 per cent, and in Estonia hardly 13 per cent. For comparison, hourly labour costs in Portugal stood at a mere 50 per cent of the average EU-15 level (see Figure 5.3).

Notes

1. We focus only on labour productivity, as statistical data for other productivity factors such as physical capital and human capital are either not available or lack comparability.
2. The Hungarian National Bank published corrected FDI data including reinvested profits (estimated at €2 billion per year for 1995–2003), but even these new data show a recent decline).
3. Sources of data are the country studies (http://www.iwh-halle.de/projects/productivity-gap).
4. Double-accounting firms; the total number of all FIEs was 26,645 (Hungarian Statistical Office (HSO)).
5. In Estonia, only the majority foreign-owned firms with more than twenty employees are registered as FIEs, the numbers are hence biased downwards.

6

An Econometric Study of Estonia and Slovenia: The Effect of FDI on Labour Productivity

Pritt Vahter

The aim of Chapter 6 is to study the effects of FDI on labour productivity in Estonia and Slovenia in the of manufacturing sector. This selection of countries allows the analysis to use firm-level panel data at a level of aggregation and scope rarely accessible even in Western economies: the research is based on firm-level panel data of the manufacturing industries of Estonia and Slovenia from the second part of the 1990s until 2001. The analysis studies the correlation between foreign equity participation in the firm and the firm's own productivity – i.e. the 'own-firm' effect in the terminological tradition of Aitken and Harrison (1999). This helps to provides the answer as to whether there exist *intra-industry* (within the same sector) spillovers from foreign affiliates to firms with no FDI and to other foreign affiliates in these two countries. The analysis also focuses on the issue whether 'own-firm' productivity effects depend on the type of FDI. More specifically: is there a difference in 'own-firm' effects between export oriented and domestic market oriented FDI? The exporting/local market orientation dimension is usually neglected in the analysis of effects of FDI on productivity (except, e.g., Kokko, Tansini and Zejan 2001; Harris and Robinson, 2001; Sgard 2001). Yet, the effects of these two types of FDI on the host economy may be quite different. This distinction is particularly relevant for the debate on how governments should design their policies to attract FDI and whether export oriented FDI is preferable for the host economy, as the policy literature sometimes assumes (e.g. *World Investment Report*, 2002). This distinguishes this analysis from the bulk of other literature, makes it particularly valuable and the interpretation of its results particularly trustworthy.

Introduction

The two countries selected for the analysis have had different stages of development and hence substantially different effects of FDI on their economies. Slovenia has the highest gross domestic product (GDP) *per capita* among the CEE transition economies. In Estonia, the level of GDP per capita is lower, but the inward FDI penetration rates have been far higher. In Estonia also the attitude to FDI, government policies and privatisation methods have been much more FDI-friendly. The reasons why investors choose the host country are therefore different for Estonia and Slovenia.

Employing panel data techniques we can account for the firm-specific time invariant effects and also for any sample selection bias. Another important issue mentioned by several authors is the non-random selection of FDI recipients. The most productive local firms may receive FDI and unless we account for this the positive productivity effects of FDI may be overestimated. In addition to the usual methods of econometric panel data, we therefore also employ a two-step procedure to correct for any sample selection bias.

This study of the 'own-firm' and horizontal spillover effects of FDI on productivity endeavours to contribute to a rapidly growing literature; it has the benefit of adding an export/local market orientation dimension to the analysis and using enterprise-level panel data for two different CEECs. One interesting finding in this section is that in Estonia the export oriented FIEs have on average a much lower labour productivity level than the domestic market oriented foreign affiliates. In Slovenia, however, the preference for export orientation of a foreign affiliate is not correlated with lower labour productivity. We detect no horizontal spillovers of FDI to domestic firms in Estonia. In the case of Slovenia, positive spillovers to DEs were found, and the spillovers to other foreign affiliates in Slovenia were not significantly different from spillovers to DEs.

Theoretical background

In order for FDI to materialise, the MNEs must possess some *firm-specific competitive advantages* that allow them to compete successfully in the foreign environment. These advantages – the firm-specific assets – can constitute production technologies, but they may also be related to special skills in management, distribution, product design, marketing and other links in the value chain, or be made up of brand names and trademarks (Caves, 1996). One can argue that, in the case of export oriented FDI, a significant part of the firm-specific advantages of a foreign firm is made up of networks, relations or other export-related know-how. The theory of FDI stresses the positive links between firm-specific knowledge-based assets and the decision to invest abroad (e.g. 1974, 1996; Dunning, 1988 Caves,). These firm-specific assets have some characteristics of a public good and can be transferred at low cost between the subsidiary of the MNE and its parent company.

Technology transfer by FDI could result in 'own-firm' and spillover effects on host economies:

(1) The 'own-firm' effect, – i.e. the average performance characteristics of foreign enterprises differ from those of the DE in the host country (and are presumably better than these of the DEs).
(2) Various spillover effects from the presence of foreign firms affect the performance of DEs (and other foreign affiliates active in the host country; spillovers are also usually presumed to be positive, at least for the DEs) (Aitken and Harrison, 1999 Blomström and Kokko, 1996; Smarzynska-Javorcik, 2004).

The extent of technology transfer to a local affiliate depends on the reasons why FDI was made in the country (host country advantages), and what role (and probably also what extent of autonomy) the local FIEs have in the MNE's value added channel. If the main reasons for investment were the low cost level of the host economy, including cheapness of labour or other factors of production, then it is less likely that higher value adding-related activities will be transferred to a local FIE. Thus the 'own-firm' or 'own-plant' effect of FDI depends on the international competitive advantage of the host country and the reasons why FDI was undertaken by this particular MNE. Higher value creating activities (e.g. the use of the results of R&D) are more likely to be allocated to a local FIE if there exists a high enough level of absorptive capacity in the local firm and/or host economy as a whole (e.g. Cohen and Levinthal, 1989; Damijan *et al.* 2003).

The advantages of FDI that presumably result in better performance (including productivity) of FDI affiliates, if compared to DEs, are well documented in the literature (see e.g. Aitken and Harrison, 1999; Görg and Strobl, 2001 Blomström and Kokko, 2003a; Smarzynska Javorcik 2004). The well-known paper by Aitken and Harrison (1999) summarises the most important reasons why economists usually assume that foreign owned-firms will have higher productivity than the rest

(Aitken and Harrison, 1999 p: 605). Firstly, superior (and possibly newer) production equipment can be transferred from the parent company to its FDI affiliate. Secondly, the foreign affiliate also receives an inflow of non-tangible assets from its parent–in the form of technological know-how, management and marketing capa bili ties, trade contracts, a coordinated network of relationships with suppliers and customers abroad, etc. This can all, assuming that the local affiliate has sufficient absorptive capabilities to use this know-how, give them significant competitive advantages over DEs. Oulton (1998: 122, 144) also argues that foreign enterprises may enjoy a lower cost of capital as they are not constrained to borrow from the local financial system. The possible inability of DEs to borrow cheaply from abroad may reduce their ability to invest in superior technology (Oulton, 1998: 144, Harris and Robinson, 2001: 4).

The overwhelming majority of authors stress the positive 'own-firm' effects of FDI. However, one may find also literature indicating the possibility that some FIEs have lower productivity than DEs. According to Harris and Robinson (2001: 4) foreign-owned plants may have lower productivity levels (at least in the short run), caused by the time lag in assimilating new plants into the FDI network. This may be caused by the significant cultural differences between the host and home countries or also by hostile policies of the host country governments towards FDI.

The usual assumption is also that MNEs are more prone to acquire local companies with higher than average productivity (Aitken and Harrison, 1999: 606 Damijan *et al.* 2003;). Reasons why FIEs may sometimes have even lower productivity levels than DEs include the nature and type of activity undertaken in the foreign-owned plant (Harris and Robinson, 2001: 5). Foreign firms may keep most of their high value added operations at home (e.g. R&D), concentrating lower value added assembly operations in the host country (due to cost and labour quality differences for example). Thus the use of lower-skilled workers and the use of possibly inferior or older technology will contribute to potentially lower productivity. This practice, although not a general one, is consistent with some empirical evidence from Japanese greenfield investments in the United States (Okamoto, 1999).

There is also an expanding literature that links exporting and productivity (Görg and Strobl; 2001, Gestrin 2001; Bernard and Jensen 1999; Delgado, Farinas, and Ruano, 2001). The causality can, as in the case of FDI and productivity, run both ways. There can exist both a 'learning-by-exporting' effect, meaning that exporting causes higher productivity of the firm, a 'self-selection' effect. 'Self-selection' means that firms with higher than average productivity are more likely to become exporters. Empirical work, for example on the United States or Western European countries, often suggests that the productivity levels in exporting firms are higher than in non-exporting ones (Bernard and Jensen Farinas and Ruano 1999; Delgado, 2001: 397). This is part of the reason why export oriented FDI is generally considered to be better for the host country than non-export oriented FDI (Gestrin, 2001: 2).

The predominant conclusion from the theoretical literature, however, is that the 'own-firm' effect of FDI on productivity is expected to be positive.

The presence of an MNE in a host country can lead to technology transfer to Des – i.e. to spillovers of FDI to local enterprises (Aitken and Harrison, 1999: 605). If foreign firms introduce new products and/or processes in their affiliates in a host country, and other FIEs may benefit from an accelerated diffusion of new technology. Spillovers are said to take place as MNEs – due to the public good characteristics of their firm-specific assets and due to these assets being at least to a certain extent non-excludable and non-rival goods – cannot reap all the benefits of their activities in a foreign location (Caves, 1996: 185).

Usually, both the theoretical and the empirical literature discuss the spillovers to domestic capital-based enterprises. However, in addition to the technology transfer to the MNE's subsidiary and the productivity spillovers to the DEs, other affiliates of MNEs in the host country may also receive part of the FDI productivity spillovers. If the share of FIEs is relatively large in the host economy, these other foreign affiliates should not be left out of the analysis as possible recipients of spillover effects. It can be argued that the external effects on other MNE affiliates can be increasingly important as the share of inward FDI in the host economy grows. Indeed, the productive knowledge that is transferred via FDI may not be totally a public good. It could sometimes be to some extent an excludable and complementary good – i.e. more like a 'club' good benefiting most the firms with good learning capabilities and with similar backgrounds (complementary goods) enabling them to understand the context specific knowledge better. Stefano Breschi has argued that a club good has many characteristics of a public good (non-rivalry for example), but is shared by only a limited number of 'club members' and thus constitutes a public good with 'members-only' access (Breschi and Lissoni, 2001). Positive spillovers of FDI may thus sometimes benefit more the FIEs than the DEs, provided that the combinations of FIEs have more complementary characteristics among them than the combinations of FIEs with DEs.

The spillovers from inward foreign investment may be intra-industry (horizontal) or inter-industry (vertical) (Smarzynska-Javorcik, 2004). Intra-industry spillovers take place between companies in the same industry, vertical spillovers flow in the direction of suppliers and customers (backward and forward linkages) of the firm under consideration.

Based on articles by Caves (1974), Blomström and Kokko (1996), Aitken and Harrison (1999: 606–7) and Smarzynska-Javorcik (2004), we can distinguish between the following main channels for spillovers: demonstration (or imitation), competition, worker mobility and supplier exporting and upgrading effects.

The *demonstration* effect is perhaps one of the simplest examples of a spillover – for instance, the case when a local firm improves its productivity by simply observing nearby foreign firms and copying some technology used by MNE affiliates (Blomström and Kokko, 1996; *World Investment Report*, 2001: 131). In other cases diffusion of new technologies and know-how may occur through labour turnover (worker mobility effects) as employees move from FIEs to DEs.

Another type of spillover is the one that functions through *competition* between enterprises. The competition effect, unlike the demonstration and worker mobility effects that are presumably positive, can be both positive and negative (Aitken and Harrison, 1999: 607; Görg and Greenaway, 2001: 4). This is an important idea, as it significantly influences the studies on spillovers. Some kind of (competition) spillover is said to take place if the entry of an affiliate leads to more severe competition in the host economy, so that local firms are forced to use existing technology and resources more efficiently or to search for new more efficient technologies (Blomström and Kokko, 1996). This can have both positive (where a local firm manages to implement superior technologies due to the increase in competition) and negative effects on the productivity of domestic (or more generally other local) enterprises. Negative effects exist particularly in the short run (Aitken and Harrison, 1999: 607; Smarzynska-Javorcik, 2004). Negative effects are possible due to the existence of fixed costs. If imperfectly competitive firms face fixed costs of production, a foreign firm with lower marginal costs will have an incentive to increase production relative to its domestic competitors. In this environment, entering foreign enterprises producing for the local market can draw sales and the demand away from domestic firms, thus forcing them to cut their production.

The productivity of domestic firms, as shown by Aitken and Harrison (1999: 608), will fall, as they spread their fixed costs over a smaller market, forcing them back up their average cost curves. If the absolute value of this productivity decline due to diversion of demand towards the FIE is larger than the positive effect due to the transfer of technology from the MNE affiliate to domestic firms, the net productivity of DEs can decline.

A further indirect source of productivity gain may be via *export* spillovers (Kokko, Tansini and Zejan 2001; Görg and Greenaway, 2001). Görg concludes that domestic firms often learn from MNCs how to export.[1] It can be argued that productivity spillovers may be different for export oriented and domestic market oriented FDI, especially when local procurement is widespread among export oriented MNE affiliates. The 2002 *World Investment Report* (*World Investment Report*, 2002: 221–48) discusses the possibly large benefits of specifically export oriented FDI. They stress two reasons why in their opinion the targeted approach of host countries towards export oriented FDI makes sense. First, the targeted approach can help countries achieve strategic objectives related to such goals as employment, technology transfer, and cluster and export development, in line with their overall development strategies. The second reason cited is the increased competition for export oriented FDI (World Investment Report, 2002: 221). However, we would like to argue here that the spillover and 'own-firm' effects still depend largely on the type of activities transferred and that this, in turn, depends on the competitive advantages of the host country, and not only on whether the affiliate sells to domestic or international markets. One cannot agree that it is automatically true that export oriented FDI is more beneficial. Blomström and Kokko (1996: 27) have demonstrated that the countries that choose to specialise in labour-intensive processes and components' production for MNEs also have to take into account that these (export oriented) affiliates are relatively 'footloose'. They have relatively few obstacles in moving to the most favourable environment as, for example, the cost level of one host country grows. In addition to that, Gestrin (2001) has made the point that it is difficult to clearly distinguish FDI that is export oriented from FDI that is not, since this orientation can change over time.

Previous empirical literature

The important conclusion from both the theoretical and the empirical literature is that productivity spillovers are difficult to measure. As Paul

Krugman points out: 'Knowledge flows... leave no paper trail[2] by which they may be measured and tracked' (Krugman, 1991). As Görg and Strobl (2001) argue, the empirical literature tries to avoid the issue since it is difficult to address the issue of how productivity spillovers take place in reality, and it focuses instead on the simpler question whether the presence of FDI affects the productivity of DEs. One addition to this approach would be to look at all other local firms –, i.e. to consider also other MNE affiliates, not only the enterprises that have no foreign owners.

The investigation usually performed in the framework of an econometric analysis based on an estimation of the production function. Labour productivity or TFP of firms (or only domestic firms) in the host economy is regressed on a number of factors assumed to have an effect on productivity. One of these factors that is commonly used to study spillover effects is the presence of foreign firms in an economic sector or region. Another factor is the variable indicating FDI presence at the firm level (e.g. an FDI dummy that is equal to 1 it the firm has FDI, Griffith, 1999; Görg and Strobl, 2001: 724–5). The presence of FDI at the sectoral/regional level is measured by the share of FDI in assets, sales, employment, etc. The estimated econometric models in literature often use log-linear form of Cobb-Douglas production function.

Studies on the direct effects of FDI on its affiliates and spillover effects on the host economy have been made with different techniques, covering both high-income as well as developing and transition countries. Often one may find results significantly different from what one would expect based on theory or policy literature (also for the transition countries). Policy makers in the host countries of FDI often simply assume that there exist positive 'own-firm' and spillover effects of FDI (*World Investment Report*, 2001). The empirical literature, with few exceptions, usually confirms the former argument that affiliates of MNEs in a host country have on average higher productivity levels than purely DEs (e.g. Harris and Robinson, 2001: 7). The picture is, however, far more diverse if one looks at the empirical analysis of FDI spillovers.

The empirical literature, including the literature on the transition economies, shows that there is little conclusive evidence to support the view that for the host country only beneficial effects of FDI exist. There is little conclusive evidence to substantiate the incentives to attract FDI (Blomström and Kokko, 2003b; Smarzynska-Javorcik, 2004). Some surveys reveal the existence of positive spillovers, others find negative ones, while the rest find 'mixed' or not significant results (Görg and Strobl, 2001: 724, Chudnovsky, López and Rossi, 2003).

The way the research is conducted vastly influences the results obtained, and therefore the policy implications made. The findings of the literature overview by Görg and Strobl (2001: 723) underline that the results may be influenced by how the presence of MNEs defined and whether cross-section or panel data is employed. In the works where case studies and/or cross-section data were used, significant positive spillover effects relating to FDI were found. On the other hand, newer studies based on panel data that account for firm-specific time invariant effects often find also insignificant spillovers to DEs or even negative spillovers (Smarzynska-Javorcik, 2004).

Whereas the analysis of intra-industry spillovers is already well established in literature, the analysis of *vertical* spillovers is quite a new field with one of the most influential papers written by Beata Smarzynska from World Bank (as working paper in 2002 and as an article in the *American Economic Review in 2004*). One result of Smarzynska-Javorcik, 2004 this paper that is especially interesting for the analysis concerning Estonia is that by using Lithuanian data she found that greater productivity benefits are associated with the domestic market rather than export oriented foreign companies. Relatively similar results are presented by Kokko *et al.* (2001) on Uruguay.

In recent literature (Driffield and Love, 2003; or e.g. Liu and Wei, 2003) the so-called 'reverse spillovers' are discussed – i.e. spillovers from DEs to FIEs. One issue that has been totally neglected in the literature on the effects of FDI on productivity is the influence of *transfer pricing* on FDI-related productivity effects: this area deserves further theoretical analysis. To our knowledge there is a lack of work that systemalically connects these two fields. Transfer pricing may have some importance where relatively large differences exist in taxes between the host and the home countries for FDI. Transfer pricing is probably not a problem when looking at spillovers, but in a productivity comparison of foreign and domestic firms it may be: it may affect the analysis of productivity differences in FIEs and DEs (and between export oriented and domestic market oriented FIEs). Sales by foreign subsidiaries, especially for export oriented firms, are often intra-MNE transactions. The values for sales and value added by foreign affiliates may be manipulated by the MNE in order to minimize its tax liabilities and are thus likely to be different from figures based on market transactions.

Data and descriptive statistics

Slovenia is more developed as a transition economy than Estonia, as is evident from a comparison of the GDP levels of the two countries. GDP per capita for Slovenia, according to the *Transition Report Update* of the European Bank for Reconstruction and Development (EBRD, 2004), was estimated to be US$ 13,851 in 2003. The corresponding level for Estonia US$ 6,120. These facts are supported by the investment development path theory of Dunning and Narula (1995): in addition to a higher GDP per capita, Slovenia also began to invest abroad long before Estonia and had a very different track record of internationalisation (Varblane *et al.*, 2001: 18–19; Rojec and Svetličič, 2003). Estonia and Slovenia also adopted different privatisation strategies and had different attitudes and policies towards FDI: Estonia has been much more FDI-friendly than Slovenia (see, e.g., the *'Index of Economic Freedom,'* 2003). based on these differences, one can then argue that the effects of FDI on productivity differ significantly between Estonia and Slovenia, and so studying these two countries can offer interesting results and policy implications.

If one takes a look at the general FDI inflow data for the economy, one can see that FDI does indeed have a stronger role in Estonia. Inward FDI stocks reached 65.9% of GDP in Estonia and only 23.1% of GDP in Slovenia in 2002 (*World Investment Report*, 2003). The FDI stock in Slovenia was US$3.209 billion in 2001 and US$5.074 billion in 2002. The corresponding figures for Estonia were US$3.160 billion in 2001 and US$4.226 billion in 2002. If compared to the ratio of FDI stocks to GDP these figures also illustrate the significant differences between the GDP level (and per capita GDP) of Estonia and Slovenia. Figure 7.1 shows the inflows of FDI in both countries.

Government policy has been traditionally much more FDI-friendly in the case of Estonia. The corporate income tax on reinvested earnings in Estonia is deterred for example, although this applies to all firms, not only FIEs. In 2002, there was a huge jump in FDI inflows into Slovenia, to a truly unprecedented level–almost €2 billion. Much of that was accounted for by the take-over of *Lek*, a blue-chip pharmaceuticals enterprise (Slovenia, 2004: 8). The majority of FDI inflows to both Estonia and Slovenia originate from the neighbouring Western European countries.

We now a look at whether the 'own-firm' productivity effects depend on the type of FDI – i.e if there is a difference in the 'own-firm' effects of export oriented and domestic market oriented FDI. We distinguish

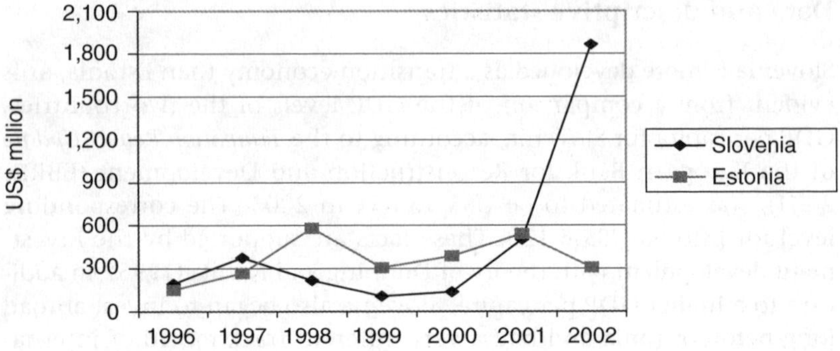

Figure 6.1 FDI inflows into Estonia and Slovenia, 1998–2002
Source: *World Investment Report* (2003).

between two dimensions: ownership, market (abroad, local): *DE* denotes a DE ($DUMF_{ijt} = 0$[3]); *FIE* denotes a FIE ($DUMF_{ijt} = 1$); *DM* denotes domestic market orientation ($DUMEXP_{ijt} = 0$); *FM* denotes foreign market orientation ($DUMEXP_{ijt} = 1$). Based on these two dimensions, we distinguish between four types of firms: domestic market oriented DEs ($DUMF_{ijt} = 0$, $DUMEXP_{ijt} = 0$); foreign market oriented DEs ($DUMF_{ijt} = 0$, $DUMEXP_{ijt} = 1$); domestic market oriented FIEs ($DUMF_{ijt} = 1$, $DUMEXP_{ijt} = 0$); and foreign market oriented FIEs ($DUMF_{ijt} = 1$, $DUMEXP_{ijt} = 1$).

Enterprise-level panel data from the Statistical Office of Slovenia and the Statistical Office of Estonia on manufacturing industries are used to study the productivity effects of FDI. For Estonia, the (balanced) panel consisted of yearly data of 326 firms over the period 1996–2001. The initial number of enterprises in the panel was 382; over fifty firms were excluded for the purposes of econometric analysis, since these firms either did not exist during the whole period of (fewer than 10% of firms) or their field of activity was not manufacturing for the whole period. According to Olley and Pakes (1996: 1265) a traditional way of accounting for entry and exit when using firm-level data is to construct a 'balanced' panel, keeping only those firms that operate throughout the entire sample period, and then compute either the OLS or some other more suitable estimator of the production function coefficients for the panel data. However, we note that this approach may also have deficiencies, as the firms that operate over the whole period are the relatively successful ones. The least successful firms that went bankrupt are left out of this analysis. However, the number of such firms excluded from the analysis is relatively small, and so the impact on the results is minor.

The Slovenian panel was significantly larger, cove yearly data for 1994–2000 for 982 firms in manufacturing industry. In addition to the standard financial statement data, the data sets contain information as to whether foreign capital has been invested in each firm. However, the definitions of an FIE and a differ for the Estonia and Slovenia data sets. For Slovenia, the usual definition of FDI-recipient firms by the OECD, the IMF or the World Bank is used. FDI 'recipient firms' are defined as firms with a foreign share equal to at least 10% of the ordinary shares or voting power (IMF, 2001: 23). For Estonia, one cannot use the 10% level for all years: owing Due to the lack of data it was not possible to calculate the share of FDI in ordinary shares or voting power for 2000 and 2001. The FDI majority share dummy variable, available from the data base of the Statistical Office of Estonia, is applied. The FDI dummy variable calculated for the 10% level would have been much more beneficial for the analysis since FDI smaller than the majority shares can still influence the performance of a firm to a significant extent. However, as annual surveys of FDI 'Foreign Investors' by the Estonian Investment Agency and Tartu University have indicated, there are relatively few firms with FDI in Estonia that have a foreign share below 50% (Varblane, 2001). However, in the case of Slovenia, there is a significant number of FIEs with foreign share between 10 and 50%.

Table 6.1 presents the number of FIEs and their share in the total number of firms over the range of the two data sets for the period. Both countries have a growing FDI share in the number of firms in the sample. In the case of Estonia, the FDI penetration rate is for all years about twice that in Slovenia. In 2000, the share of FIEs in the total number of enterprises was 23.3% in the Estonian sample and 12.8% in the Slovenian sample[4]. In Slovenia, inward FDI is far less spread throught the economy than in Estonia. The penetration of FIEs, measured by various indicators such as employment, sales or value added, is also lower for Slovenia.

The analysis of the descriptive statistics of the Slovenian panel of enterprises active in the manufacturing sector shows that the share of FIEs in the number of firms was 12.8% in 2000. These firms accounted for 33.2% of sales, 38.2% of exports, 18.7% of employment, 21.9% of value added and 24.1% of tangible fixed assets. In the Estonian panel of the manufacturing industry firms, the share of FIEs in the number of firms was 26.1%; these firms accounted for 47% of sales, 58.5% of

Table 6.1 Data description, 1994–2001

Year	No. of FIEs			FIEs' share in no. of firms	
	Estonia[a]	Slovenia	Slovenia[a]	Estonia[a] (%)	Slovenia (%)
1994	–	91	54	–	9.3
1996	69	101	73	21.2	10.3
2000	76	126	91	23.3	12.8
2001	85	–	–	26.1	–

Notes: *a* Majority-owned FIEs.
– Not available.
Source: Own calculations based on panel databases of Slovenian and Estonian enterprises in manufacturing industries.

exports, 39.4% of employment and 42.5% of R&D costs. As in Slovenia, the in Estonia FIEs are larger than Des, and indeed more export oriented. They also tend to spend more on R&D per enterprise than DEs. Smaller FIEs are quite common in Estonia, whereas in Slovenia FDI has been concentrated in a relatively small number of large enterprises.

In Table 6.2 the descriptive statistics on productivity in the manufacturing sectors of Estonia and Slovenia are presented, based on the enterprise-level panel data. The tables describe differences between productivity levels of four types of firms included in this study. Also information on capital-labour ratio is included. Labour productivity is measured as sales per employee or value added per employee. In the case of Estonia, it is also interesting to take a look at the data, from whose analysis a very large foreign affiliate of Elcoteq (a well known

Table 6.2 Descriptive statistics on the share of FDI in Estonia and Slovenia, 2001 and 2000 per cent

Year / FDI share in	Estonia 2001	Slovenia 2000
Sales	46.6	33.2
Exports	58.5	38.2
Employment	39.4	18.1
Value added	48.7	21.9
Tangible fixed assets	54.4	24.1
No. of firms	26.1	12.8

Source: Own calculations, based on the enterprise-level panel databases of Estonian and Slovenian manufacturing.

Table 6.3 Estonia: productivity differences between the four types of firms, 1996–2001, 000 kroons

Estonia, 000 kroons			**Year**						**Growth (2001/ 1996) (%)**
DUMM`	**DUMEXP**	**Data**	**1996**	**1997**	**1998**	**1999**	**2000**	**2001**	
0	0	Productivity (sales: empl.)	298	398	447	436	499	538	80.7
0	1	Productivity (sales: empl.)	217	275	264	277	350	425	96.3
1	0	Productivity (sales: empl.)	623	739	884	802	944	1116	79.0
1	1	Productivity (sales: empl.)	335	399	388	374	452	532	59.0
Total of DE		Productivity (sales: empl.)	253	327	350	349	415	479	89.2
Total of FIE		Productivity (sales: empl.)	407	480	497	475	542	644	58.1
(FIE/DE) ratio of productivity (sales: empl.)			1.609	1.466	1.420	1.359	1.305	1.344	–16.5
(FIE/DE) ratio of capital: labour ratio			4.092	3.014	2.310	2.039	1.812	1.834	–55.2
DUMEXP 0		Productivity (sales: empl.)	341	447	510	507	577	658	93.0
DUMEXP 1		Productivity (sales: empl.)	249	311	307	315	393	479	92.6
(*DUMEXP* 1: *DUMEXP* 0) ratio of productivity (sales: empl.)			0.730	0.695	0.602	0.621	0.682	0.728	–0.2
Without Elcoteq:									
1	1	Productivity (sales: empl.)	366	458	435	426	522	588	60.6

Notes: *DUMM* denotes the FDI dummy (= 1, if at least 50% of voting power belongs to the foreign investor, else = 0); *DUMEXP* denotes the export orientation dummy (= 1, if the share of exports in sales is at least 50%, else = 0; FIE = Foreign investment enterprise; DE = Domestic enterprise; empl. = No. of employees; Capital: labour ratio = Tangible fixed assets per employee.
Source: Own calculations, based on enterprise-level panel database of Estonian manufacturing.

foreign electronics company active in Estonia, among the largest firms in Estonian manufacturing) has been excluded.

The statistics in Table 6.3 show that foreign affiliates have significantly higher labour productivity in Estonia than DEs. This result also holds for the value added-based approach to labour productivity measurement (Vahter, 2004). In 2001, the labour productivity of DEs based on sales per employee in Estonian manufacturing was (from on the panel of 326 enterprises) on average 479,040 Estonian kroons. In FIEs, however, the same figure amounted to 643,890 kroons – i.e. it was 34% higher than the labour productivity level of DEs. In 1996, the corresponding ratio of FIEs to DEs was higher than in 2001 – FIE productivity surpassed that of DEs by 61%.

The big difference in productivity between those two types of enterprises is also seen when the value added per employee is studied. These results indicate that the FIE level surpasses the DE level by almost 50%; in 2001 the FIE: DE ratio was 1.457 and in 1996 it was 1.410. This significant difference in productivity levels is to a large extent caused by the fact that FIEs employ more capital per employee than local domestic capital-based firms. The gap in the capital: labour ratio is even larger than in the productivity of labour. The FIE: DE ratio of the capital: labour ratio was 1.834 in 2001 and at the beginning of the studied period (in 1996) FIEs used four times more capital per employee in production than DEs. This FIE: DE ratio has, however, fallen over the years as the capital: labour ratio of DEs has, due to investments in physical capital, grown rapidly (122% over the period 1996–2001), whereas that of FIEs has stayed roughly the same.

In the case of Slovenia, as in Estonia, we can see that the labour productivity level of FIEs is on average much higher than that of DEs; in 2000, it was 2.25 times higher. In Estonia in 2000, in contrasrt, it was only 1.34 times higher. The difference between these two types of firms is this much larger in the case of Slovenia than in the case of Estonia. The capital: labour ratio of FIEs in Slovenia generally surpasses that of DEs.

Among the four types of firms, in Estonia in 2001 the lowest productivity was found in Des that produce predominantly for export markets- sales per employee were 425,090 kroons, value added per employee was 98,750 kroons. The ranking of the four types of firms under consideration in Estonian manufacturing in 2001 (from the group with the highest level of the indicator to the one with the lowest), based on sales per employee as a measure of labour productivity, proved to be as follows: first, came domestic market oriented FIEs;

second, domestic market oriented DEs; third, foreign market oriented FIEs; and fourth foreign market oriented DEs.

The highest labour productivity is to be found in domestic market oriented FIEs: 1,115,550 kroons in 2001. This ranking of second and third place changes some what over the period but the first and the last places remain the same. The main conclusion from Table 7.3 is that in Estonia the export orientation of a firm is correlated with lower labour productivity. In 2001 and 1996, the labour productivity of export oriented enterprises was about 73% of the corresponding level of domestic market oriented firms. What can also be inferred from these results is that the aim of FDI in the manufacturing sector, except for local market oriented FDI, has been to benefit from Estonia low labour costs.

Former empirical studies in Estonia (see, e.g., Hannula and Tamm, 2003) have stressed that FIEs have on average much higher labour productivity levels than DEs. Now it is possible, based on this analysis, to see that there is a productivity difference in Estonia due only to domestic market oriented firms, whereas export oriented foreign affiliates have indicators more than twice as low (in the case of sales per employee, e.g. in 2001) as domestic market oriented firms with FDI. The productivity level of the export oriented FIEs is comparable to that of the domestic market oriented DEs. In the period 1998–2001, it is below that already relatively low level.

We have also computed the productivity indicator for export oriented foreign affiliates in Estonia without the electronics manufacturing services provider Elcoteq (Table 7.3) (a big company: in some years with more than 3000 employees). The reader can see that without Elcoteq the indicators of export oriented FIEs are much higher than with it. Tabour productivity as sales per employee, without Elcoteq, is 588,160 kroons for 2001 in this group of firms. Before the exclusion of Elcoteq from our sample, the corresponding figure amounted to 532,200 kroons. This means that Elcoteq, with relatively low productivity has, due to its size, significant impact on the analysis of productivity in our framework.

Let us now turn to Slovenia. The results for this transition country are given in Table 6.4, and one can see significant differences from Estonia. The rankings of firms by type differs also between the period 1994–8 and 1999–2000. In the former, the ranking, starting from the group of firms with the highest labour productivity (sales per employee) is as follows: first, foreign market oriented FIEs; second, domestic market oriented FIEs; third, domestic market oriented DEs; and fourth, foreign market oriented DEs.

In 1999, the first and second group changed their position (note also the differences in sales per employee in 1999 and 2000 from Table 7.4): first, domestic market oriented FIEs; second, foreign market oriented FIEs; third, domestic market oriented DEs; and fourth foreign market oriented DEs.

Based on these figures, the conclusion is that export orientation in Slovenia – unlike in Estonia – is not associated with lower labour productivity levels. Export oriented and domestic market oriented firms have on average about the same level of productivity. If the years 1999, and especially 2000, with peculiarly high indicators are excluded, we can conclude that export oriented FIEs have the highest level of productivity among the four types of firms. Quite similar levels (in 1999 and 2000 also higher) are found for the domestic market oriented FIEs. The DEs, regardless of whether they are export oriented or domestic market oriented, lag far behind. The export oriented DEs have, however, the lowest productivity level among all the firms. In any analysis of the results from Table 6.4, some caution is advised when discussing the implications of the 2000 results of 2000. The big leap in productivity level of the first-round group in 2000 (domestic market oriented foreign firms) can be attributed to the small number of firms (with minority foreign ownership) and to a possible measurement error in the case of these firms. If we look only at FIEs with a majority foreign share, there is not that significant a growth in productivity of export oriented FIEs in 2000.

The results of the ranking of export oriented FIEs for Slovenia (Table 6.4) and for Estonia (Table 6.3) are in sharp contrast. The reasons for Slovenia having this group of enterprises as a top performer and Estonia having it as a low productivity group may to a large extent be the result of the different location-specific advantages that these two countries provide for investors. Based on the information from investor motivation surveys of Estonia and Slovenia, we find that there are large differences in main motives of FDI between the two countries (e.g. *Foreign Investor*, 2000; OECD 2002: 14). In Estonia, the relatively low production costs, including labour costs, have been a predominant factor affecting investment decisions (Varblane, 2001).

Surveys on the motivation of foreign investors investing in Slovenia, on the other hand, show that as far as labour is concerned it is clearly the quality and not the cost that attracts foreign investors (OECD, 2003). In Slovenia, only 1.8% of foreign investors emphasise the motive of low labour cost; however, quality of labour is a motive for 26.9% of the FIEs (OECD, 2003: 14). This is not surprising, as labour

Table 6.4 Slovenia: productivity differences between the four types of firms, 1994–2000, 000 tolars

Slovenia, 000 tolars			**Year**							**Growth (2000/ 1994)**
DUMF	**DUMEXP**	**Data**	**1994**	**1995**	**1996**	**1997**	**1998**	**1999**	**2000**	**(%)**
0	0	Productivity (sales: empl.)	6,759	7,553	8,724	10,509	10,883	11,639	13,321	97.1
0	1	Productivity (sales: empl.)	5,498	6,065	7,044	8,356	9,222	9,483	11,779	114.2
1	0	Productivity (sales: empl.)	11,849	14,702	16,198	18,420	20,522	22,633	42,820	261.4
1	1	Productivity (sales: empl.)	13,985	15,137	18,683	20,456	22,941	21,987	25,074	79.3
Total of DE		Productivity (sales: empl.)	6,011	6,643	7,672	9,096	9,827	10,296	12,349	105.4
Total of FIE		Productivity (sales: empl.)	13,339	15,025	18,079	20,071	22,513	22,118	27,793	108.4
(FIE/DE) ratio of productivity (sales: empl.)			2.219	2.262	2.357	2.207	2.291	2.148	2.251	1.4
DUMEXP 0		Productivity (sales: empl.)	7,181	8,106	9,315	11,107	11,622	12,831	15,794	119.9
DUMEXP 1		Productivity (sales: empl.)	6,562	7,264	8,641	10,236	11,711	12,279	14,821	125.9
(*DUMEXP* 1: *DUMEXP* 0) ratio of productivity (sales: empl.)			0.914	0.896	0.928	0.922	1.008	0.957	0.938	2.7
(FIE/DE) ratio of capital: labour ratio			1.623	1.535	1.671	1.723	1.567	1.423	1.441	–11.2

Notes: *DUMF* denotes the FDI dummy (= 1, if at least 10% of voting power belongs to the foreign investor, else = 0); *DUMEXP* denotes the export orientation dummy (= 1, if the share of exports in sales is at least 50%, else = 0; FIE = Foreign investment enterprise; DE = Domestic enterprise; empl. = No. of employees; capital: labour ratio = Tangible fixed assets per employee.
Source: Own calculations, based on enterprise-level panel data of Slovenian manufacturing.

Table 6.5 Average monthly wages in manufacturing, Selected CEECs, 2000–2

Country	2000	2001	2002
Estonia (EER)	309.6	341.1	376.1
Slovenia (EER)	763.1	820.0	868.0
Ratio SLO: EST	2.47	2.40	2.31

Sources: Statistical Office of Estonia, Statistical Office of Slovenia.

costs in Slovenia are the highest among the transition countries in Central and *Eastern Europe*. In 2002 the average monthly wages in Slovenian manufacturing were 2.3 times higher than in Estonia (see Table 7.5).

The surveys for Estonia also indicate that export oriented investors have a different motivation for investing in Estonia than domestic market oriented investors. Exporters are more motivated by the costs of production and the labour force than by the market potential, as they do not plan to supply the domestic market. The non-exporters, in turn, are more motivated in tapping the new market and benefiting from the expected market growth. Exporters represent mainly the chemical, wood processing and furniture, electronics, textile, machinery and engineering industries, while non-exporters are mainly from the food and beverage and construction material industries (Varblane and Ziazic, 1999).

General model and econometric concerns

In order to examine the effects of FDI on productivity, we follow the general model (production function approach) of the literature (as specified in Aitken and Harrison, 1999, with some added features. One difference in our study is that the inputs and the dependent variable are given per employee. i.e. the dependent variable is not output as (as in Aitken and Harrison, 1999) but a measure of *labour productivity*, sales per number of employees. Input variables this include the capital: labour ratio, materials per employee, etc. The export orientation dummy variable $DUMEXP_{ijt}$ is also included in order to account for export oriented firms. The following model is estimated:

$$Y_{ijt} = C + \beta_1 DUMF_{ijt} + \beta_2 DUMEXP_{ijt} + \beta_3 DUMEXP_{ijt} \cdot DUMF_{ijt} + \beta_4 FDI\beta sector_{ijt} + \beta_5 DUMF_{ijt} \cdot FDI\_sector_{ijt} + \beta_6 X_{ijt} + \beta_7 Z_{jt} + \varepsilon_{ijt}$$

The log of sales per number of employees, $Y_{ijt} = log(sales_{ijt} / employees_{ijt})$, for firm i in sector j at time t (deflated by the producer price index, PPI) is regressed on the vector of inputs/control variables per employee (X_{ijt}, these are given all in logarithms), sector dummies Z_{jt}, an export orientation dummy and its interaction dummy with the measure of FDI and measures of foreign ownership $DUMF_{ijt}$ and FDI_sector_{ijt}. The vector of control variables X_{ijt} includes variables $LNTFA_{ijt}$, $LNLABC_{ijt}$ *and* $LNMATER_{ijt}$, $DUMINT_{ijt}$ (for Slovenia) or $DUMRD_{ijt}$ (for Estonia). These variables used in the regression analysis are defined below. C is a constant and ε_{ijt} is the error term.

$DUMF_{ijt}$ indicates an FDI dummy variable. This variable identifies whether or not a firm has FDI (the threshold level is 10% of voting power in the firm for Slovenia and 50% for Estonia); $DUMF_{ijt} = 1$ if the firm is an FIE, $DUMF_{ijt} = 0$ if it is a DE. If foreign ownership in a firm increases that firm's productivity, we should observe a positive coefficient for $DUMF_{ijt}$. The variable $DUMEXP_{ijt}$ is the export orientation dummy. It takes the value of 1 if a firm has the share of its exports in its sales at least as high as 50%, and the value of 0 otherwise. As exporting may have a positive effect on labour productivity, we expect this variable to have a positive coefficient. The interaction dummy between $DUMF_{ijt}$ and $DUMEXP_{ijt}$ in order to capture interaction effects is $DUMEXP_{ijt} \cdot DUMF_{ijt}$. It allows us, together with the variables $DUMEXP_{ijt}$ and $DUMF_{ijt}$, to distinguish between the four types of enterprises. Where export oriented FIEs have higher labour productivity level than the domestic market oriented FIEs, the coefficient of this variable will be positive.

FDI_sector_{ijt} is the share of FDI in a sector as measured by the ratio the sum of the assets of the FIEss in a sector (with each FIE's own assets subtracted) to the sum of the assets of all the firms in the sector.[5] Sectors are defined at the NACE double-digit level. This indicator is used for measuring horizontal spillover effects,. if the productivity advantages of foreign capital spill over to domestic firms in the same sector, the coefficient of this variable should be positive.

The coefficient on the interaction between firm-level and sector-level FDI is captured by $DUMF_{ijt} \cdot FDI\_sector_{ijt}$. It allows us to determine if the effects of a foreign presence on other foreign firms differ from the effects on domestic firms. $LNTFA_{ijt}$ is the log of the tangible fixed assets per employee, a proxy for the logarithm of $(K{:}L)$ ratio. A proxy for the skill intensity of the employees of the firm is also included and is measured by $LNLABC_{ijt}$, the log of the labour costs per employee. As the dependent variable is based on sales, the right-hand side of the equation must take account of materials: $LNMATER_{ijt}$ is the log of materials per employee. The dummy variable $DUMINT_{ijt}$ takes the value of 1 if the firm has intangible fixed assets, 0 otherwise. An alternative to this variable is $DUMRD_{ijt}$; it is equal to 1 if the firm has R&D expenditures.

Sector dummy variables are also used in the regression model in order to capture sector specific effects and year dummy variables are used in order to account for any trend effects. The inverse of Mill's ratio (seen. 6) is employed in a Heckman-type two-step procedure in order to account for any sample selection bias in estimation (see P. 10)

The use of panel data has several benefits over the usual cross-section data (see, e.g., Chapter 13 in Greene, 1993; Wooldridge, 2002). By using panel data it is possible to account for the individual heterogeneity of objects in the analysis (e.g. the absorptive capacities of the firms, etc.). The easiest way to account for heterogeneity would be for example, including a separate dummy variable for each object in the model. Secondly, panel data give more information on data, more variety, less collinearity between variables, many more degrees of freedom and better efficiency of estimators (Chapter 13 in Greene, 1993; Baltagi, 2001).

Some econometric concerns need to be addressed before estimating the general model of our study. The first is the choice of the method for estimation, based on the panel data for Estonia and Slovenia: whether one should use the simple pooled least squares model (pooled LS) or a RE or FE model. Pooled LS has a multitude of disadvantages when panel data is used. Pooled LS does not take into account the time invariant firm-specific effects that are likely to exist if the researcher employs panel data. Not taking these effects into account (if they exist) – i.e. just running OLS for pooled data – would lead to biased and inconsistent estimation results. The common remedy could be to use RE or FE models instead. These both include object-specific time invariant effects but have different assumptions on the essence of these effects. The FE model assumes that differences across units can be captured in differences in the constant term. The FE model is a reasonable approach when the researcher can be confident that the differences between firms can be viewed as parametric shifts of the regression function (Greene, 1993: 466). In the case an RE effects model, individual-/firm-specific constant terms are viewed as randomly distributed across cross-sectional units (Greene, 1993: 469). The inevitable question is: which approach should be used? There are different views; Mundlak (1978), for example, argues that one should always treat individual effects as random (Greene, 1993: 479). On the other hand, FE models have considerable advantages over RE models, as FE models (where individual effects are indeed correlated with other regressors, unlike the assumption of the RE model) may suffer from the inconsistency due to omitted variables (Wooldridge, 2002).

One way of choosing between an RE or FE model is by looking at the panel data used in the analysis. Where we have a sample of micro data as a random draw from a population, the RE model might be appropriate. This reasoning suggests an RE model for our analysis. In addition, there is a formal aspect to the question. To test, whether an RE or FE model should be favoured, a Hausman specification test can be used (Wooldridge, 2002). When choosing between the RE or FE model, we have to bear in mind that for the FE model we cannot find the effect of these variables that are constant for the object over the panel range (in our case, possibly the sector dummy variables for those enterprises that do not move between categories of FIE and DE firms and also the FDI dummy) as these are differenced out. In the case of the RE model, one can also find these effects. The implication for our analysis is that FE and RE models are different in the sense that the FE model takes into account only the dummy variables for those firms for which the values of the FDI dummy and the export orientation dummy change over the period. The RE model uses dummy variables of all firms. This means that in the case of the FE model, a substantial part of the information in the data is left unused. The FE estimator uses only the across-time variation, which tends to be much lower than the cross section one (Arnold, 2003). In Tables 6.6 and 6.7 the results both for FE and RE models will be presented.

One issue that has been mentioned by several authors is the non-random selection of FDI recipients (Djankov and Murrell, 2002 Smarzynska-Javorcik, 2002; Arnold, 2003; Damijan *et al.* 2003,). Where the most productive local firms receive FDI – unless it is accounted for – an over-estimation of the positive productivity-related effects of FDI may be the result. To take account of this possibility, after estimating the usual RE and FE models the econometric analysis continues with a Heckman-type two-step procedure in order to control for any possible sample selection bias (this is also called the Heckman model – see, e.g., Heckman, 1979).

In the first stage, the probit model is estimated. The dependent variable is the dummy variable $DUMF_{ijt}$ for an FIE (equal to 1 if the firm has foreign ownership). Independent variables that might affect the choice of the foreign investor as to whether to invest or not in the firm include labour productivity, export orientation, skill intensity (labour costs per employee) and fixed assets per employee as a proxy for the capital: labour ratio. After the estimation of this first stage, the inverse

Table 6.6 Slovenia: regression results of the estimated model, the effect of FDI on productivity, 1994–2000; RE and FE models, the dependent variable is the logarithm of labour productivity (*sales*$_{ijt}$/*employees*$_{ijt}$)

	RE model			FE model		
	Coef.	**Std. err.**	**P > |z|**	**Coef.**	**Std. err.**	**P > |t|**
LNTFA	0.1089	0.0158	0.000	0.1042	0.0141	0.000
LNLABC	0.5583	0.0761	0.000	0.5191	0.0707	0.000
DUMF	0.1191	0.0971	0.220	0.0837	0.084	0.320
DUMEXP	–0.0168	0.0262	0.522	0.0196	0.233	0.401
DUMEXP·DUMF	0.1442	0.1089	0.185	0.1839	0.114	0.107
FDI_sector	0.3417	0.1601	0.033	0.3949	0.13	0.002
DUMF·FDI_sector	–0.5287	0.7395	0.475	–0.7182	0.7449	0.335
LNMATER	0.0699	0.0159	0.000	0.0803	0.0153	0.000
DUMINT	0.0204	0.0138	0.140	0.0261	0.0126	0.038
Constant	3.0618	0.5234	0.000	3.1863	0.4905	0.000
Sector dummies		Yes			Dropped	
Year dummies		Yes			Yes	
No. of observations		**6,780**			**6,780**	

Note: Heteroscedasticity-autocorrelation robust standard errors.
Source: Own calculations, based on the panel data of Slovenian enterprises.

Table 6.7 Estonia: regression results of the estimated model, the effect of FDI on productivity, 1996–2001; RE and FE models, the dependent variable is the logarithm of labour productivity (*sales*$_{ijt}$/*employees*$_{ijt}$)

	RE model			FE model		
	Coef.	**Std. err.**	**P > |z|**	**Coef.**	**Std. err.**	**P > |t|**
LNTFA	0.0543	0.0141	0.000	0.0371	0.0155	0.017
LNLABC	0.6663	0.0475	0.000	0.6731	0.046	0.000
DUMF	0.0572	0.0685	0.404	0.0128	0.0822	0.876
DUMEXP	0.0603	0.0247	0.015	0.0767	0.022	0.001
DUMEXP·DUMF	–0.1268	0.0556	0.022	–0.1075	0.0667	0.107
FDI_sector	–0.0404	0.0766	0.598	0.0026	0.0715	0.971
DUMF·FDI_sector	0.3018	0.1086	0.005	0.3421	0.1145	0.003
LNMATER	0.3154	0.0306	0.000	0.2936	0.031	0.000
DUMRD	0.000699	0.0205	0.973	–0.0074	0.0194	0.704
Constant	1.0518	0.1603	0.000	1.231	0.173	0.000
Sector dummies		Yes			Yes	
Year dummies		Yes			Yes	
No. of observations		**1,915**			**1,915**	

Note: Heteroscedasticity-autocorrelation robust standard errors.
Source: Own calculations, based on the panel data of Estonian enterprises.

of Mill's ratio[6] (also called the non-hazard ratio) is calculated and included as a separate extra variable in the second -tage estimation in the regression function. In this second stage, the RE model is estimated (according to the general form already presented, with the inverse of Mill's ratio as an additional variable).

The variables in the probit model for estimating the probability of receiving FDI are given in n. 7.

Estimation results

The estimation results for the FE and RE model (with and without correction for sample selection bias) are given in Tables 6.6–6.9 for Estonia and Slovenia. The model selection is based on the *F*-test, the Breusch–Pagan LM-test and the Hausman-test:

(1) Pooled LS vs. FE: *F*-test.
(2) Pooled LS vs. RE: LM-test.
(3) FE vs. RE model: Hausman-test.

The following test-statistics are given for the model as specified in Tables 6.6 and 6.7. The value of the *F*-test statistic is: (a) for Estonia $F = 8.82$ ($p = 0.000$); (b) for Slovenia $F = 23.23$ ($p = 0.000$). The null hypothesis (pooled LS) is rejected for both countries in favour of the FE model. This means that there exists an unobserved heterogeneity effect. The value of the LM = statistic is: (a) for Estonia LM = 1316.72 ($p = 0.000$); (b) for Slovenia $LM = 10907.99$ ($p = 0.000$). The null hypothesis (pooled LS) is rejected for both countries in favour of the RE model. These results show again that there exists an unobserved heterogeneity effect. The Hausman-test enables us to choose between the RE and the FE model. The Hausman-test statistic is: (*a*) for Estonia $\chi^2 = 65.42$ ($p = 0.000$); (b) for Slovenia $\chi^2 = 146.99$ ($p = 0.000$). The null hypothesis (RE model) is rejected for the models of both countries; the RE model is not favoured and the FE model is favoured.

The Hausman-test indicates that we should prefer the FE model over the RE model. Due to the fact that the FE model considers only these firms that have a change in dummy variables suchs as $DUMF_{ijt}$ over the period, the RE model is also given, which considers all firms (including also those that are FIE or DE) for the whole period under consideration. The results are not qualitatively very different between the two specifications, but both models are presented as they make use of different information in data, so both could be of interest. The results of

Table 6.8 The first stage of the Heckman-type two-step procedure: the probit model, estimation of the probability of receiving FDI (dependent variable: FDI dummy)

	Probit model[a]					
	Coef.	Std. err.	P > |z|	Coef.	Std. err.	P > |z|
PROD	1.02E-05	3.68E-06	0.005	0.0004	0.0005	0.345
EXPSALES	2.123	0.3029	0.000	2.4442	0.3747	0.000
LABC	0.0006	7.56E-05	0.000	0.0239	0.0056	0.000
FAPEREMP	–2.66E-06	5.34E-06	0.619	–	–	–
TFAPEREMP	–	–	–	0.0006	0.0008	0.463
Constant	–7.5757	0.4768	0.000	–5.7691	0.4813	0.000
No. of observations		**6,810**			**1,949**	

Note: *a* Random effects probit.
Source: Own calculations, panel data of Slovenian and Estonian enterprises.

the Heckman-type two-step procedure for accounting for any sample selection bias are given in Tables 6.8 and 6.9. Note that in this case the random effects probit model over all the years of the sample is used.

Based on the estimation results of the model as in Table 6.7, but without variables $DUMEXP_{ijt}$ and the interaction variable $DUMEXP_{ijt} \cdot DUMF_{ijt}$, we find that in Slovenia foreign equity participation is positively correlated with a firm's productivity level (the 'own-firm' effect). The coefficient of the FDI dummy was positive, relatively large and significant; but after including the export orientation dummy and the interaction dummy between FDI presence in a firm and its export orientation, it proved to be positive but not significant. We test for the differences in productivity-related 'own-firm' effects between export oriented and domestic market oriented enterprises. For that purpose the coefficients of the three variables $DUMF_{ijt}$, and $DUMEXP_{ijt}$, $DUMEXP_{ijt} \cdot DUMF_{ijt}$ are studied. In order to find the difference in the productivity of export oriented FIEs from domestic market DE-level productivity, these three coefficients are added up; to find the domestic market oriented FIE effect, the coefficient of $DUMF_{ijt}$ suffices. As in our Slovenian model, these variables are not statistically significant; we cannot draw further inferences about the differences in the productivity-related 'own-firm' effects of the export and domestic market oriented FDI, but have to rely on the results presented on P.00 and 00.

For Estonia, export orientation together with the majority of foreign capital in a firm indicates, on an average, a much lower labour productivity level – which is a different result from Slovenia (see Table 7.6).

Table 6.9 The effect of FDI on productivity; the RE model including the inverse of Mill's ratio; the dependent variable is the logarithm of labour productivity ($sales_{ijt}/employees_{ijt}$)

	Slovenia, RE model			Estonia, RE model		
	Coef.	**Std. err.**	**P > |z|**	**Coef.**	**Std. err.**	**P > |z|**
LNTFA	0.1089	0.0157	0.000	0.0544	0.0142	0.000
LNLABC	0.5473	0.0759	0.000	0.6670	0.0487	0.000
DUMF	0.1144	0.0976	0.241	0.057	0.0685	0.406
DUMEXP	–0.0164	0.0262	0.531	0.0605	0.0247	0.014
DUMEXP·DUMF	0.1403	0.1083	0.195	–0.1266	0.0555	0.023
FDI_sector	0.3433	0.1601	0.032	–0.0406	0.0767	0.596
DUMF·FDI_sector	–0.5046	0.7422	0.497	0.3021	0.1085	0.005
LNMATER	0.0704	0.016	0.000	0.3154	0.0306	0.000
DUMINT	0.0195	0.0137	0.154	–	–	–
DUMRD	–	–	–	0.0007	0.021	0.975
Inverse of Mill's ratio	0.0017	0.001	0.073	–0.0034	0.1082	0.755
Constant	3.1188	0.1583	0.000	1.05	0.1641	0.000
Sector dummies		Yes			Yes	
Year dummies		Yes			Yes	
No. of observations		**6,780**			**1,915**	

Notes: Heteroscedasticity-autocorrelation robust standard errors.
– = Not available.
Source: Own calculations, panel data of Slovenian and Estonian enterprises.

This difference again shows the different competitive advantages of these two countries: while Slovenia's advantages are in the higher-value added, skilled labour and higher-productivity-related sectors, Estonia is attracting FDI more through costs lower than in the investors' home countries. The estimation results for Estonia, at least concerning the RE model, this affirm the view based on descriptive statistics from Table 7.3.

We also tested for intra-industry (horizontal) spillovers from foreign affiliates to firms with no FDI (DEs) and to other foreign affiliates. The general assumption based on the theory is that this effect is positive (where negative competition effects do not dominate).

For Slovenia:, there were positive (horizontal) spillovers from FIEs to domestic firms; the coefficient of the variable $DUMF_{ijt}{\cdot}FDI\_sector_{ijt,}$ indicating spillovers to other FIEs, was negative but proved to be not significant after correcting the standard errors for heteroscedasticity. There were positive spillovers to Des, meaning that the presence of FIEs in the manufacturing sector (at the NACE double-digit aggregation level) affects the productivity of DEs in this sector. The results stay the same if a lagged spillover variable is used. Thus the FIEs were not-better than the DEs in absorbing and benefiting from FDI productivity spillovers. This may show that the absorptive capacity of the DEs in Slovenia is relatively high. Another explanation may be that the productive knowledge transferred via FDI in Slovenia is still to a certain extent a non-excludable good and to a lesser extent a 'club' good with 'members-only' access.

The results for Estonia regarding spillovers were again, similarl to the 'own-firm' effects, different from the results for Slovenia: in fact, they were the opposite of the Slovenia results for. The spillover effect of FDI penetration in the same sector in Estonian manufacturing was not significant for DEs in the same sector. Initially, positive and relatively large significant effects for other FIEs in the same sector were found. This could a at first glance be taken to indicate that other FIEs received more positive spillovers of FDI than DEs. However, this result, unlike the rest, is not robust to different specifications of the model. Using a lagged variable for spillover analysis indicated no significant spillovers to other FIEs. These results were also tested by splitting the sample and running the regression model again only on DEs, thus naturally without any variables indicating FDI presence at the firm. The results of that approach confirmed the findings for both Slovenia and Estonia that have been presented here.

These results stay basically the same for different specifications – : for the RE and FE models and for the Heckman-type two-step procedure used for accounting for any possible sample selection bias in data. Firms with higher labour productivity (see Table 6.8) had a higher probability of receiving FDI in Slovenia, but not in Estonia.

The Mill's ratio variable that was calculated and added to the model was significant for Slovenia (at the 10% level). This suggests that there exists some sample selection bias in the case of Slovenian data-firms with high productivity in Slovenia attract more FDI. A continuous variable – the share of exports in sales – was also tried instead of an export orientation dummy. This change did not alter the basic conclusions given here. Exclusion of Elcoteq from the panel and estimating the same models again did not alter the basic qualitative results obtained here.

One consideration that had to be studied more carefully was the data for 2000 (the last year in the sample) for Slovenia. The year 2000 looks rather 'strange' in Slovenian manufacturing, as there is a Significant increase in productivity compared to 1999. This could possibly have been so due to some large merger or acquisition or measurement error. It proved to be possible to isolate the four firms that caused this 'leap' in productivity, and after excluding these four firms from the sample, the estimation of the models was performed again. The exclusion of these firms affected the values of coefficients in the regression analysis to a small extent but the qualitative interpretation of the results stayed basically the same. This big 'leap' in labour productivity levels was peculiar to only one type of enterprise in the Slovenian panel-the domestic market oriented minority foreign ownership FIEs. One reason for these effects on the analysis caused by a single firm or a few enterprises is that both countries are small economies where one big FDI in put can affect the average characteristics of firms and sectors to a significant extent.

Conclusions

FDI can be an important source for productivity growth and swifter transformation processes in the transition countries. However, FDI can theoretically cause both positive and negative spillover effects to the host economy. Our analysis of the panel data from Slovenia and Estonia in line with previous empirical studies, shows, that the growing tendency of the CEE governments to offer special incentives for FDI has relatively weak grounds. The justification (at least in the policy literature) for these incentives (in countries other than Estonia and Slovenia) has mostly been the possible beneficial effects caused by the transfer of technology from a parent company to its local affiliate and the related (positive) spillover effects to the host country.

The different stages of development in Estonia and Slovenia also imply differences in the effects of FDI to the economy. Indeed, as this study indicates, there are different consequences for productivity-related FDI effects, particularly when we also employ the export/local market dimension of the firms under analysis.

Foreign-owned firms have, on an average, higher labour productivity levels than DEs in both Estonia and Slovenia. However the results are more surprising when we divide these firms into sub-groups by their export orientation. For Estonia, the export orientation, together with the majority of foreign capital in a firm, indicates a much lower labour productivity level. This is the opposite to Slovenia. Export orientation of a FIE is not correlated with lower labour productivity and until 1998 export oriented foreign affiliates in Slovenia had significantly higher productivity even than local market oriented FIEs. This difference in the findings also shows the different competitive advantages of the two countries: whereas Slovenia's advantage is in higher-value added, skilled labour and higher-productivity-related sectors, Estonia attracts FDI more due to lower costs compared to the investors' home countries. This view is also supported by the labour cost data and investor motivation survey data from these two countries.

In this study, we also tested for the intra-industry spillovers from foreign affiliates to firms with no FDI (DEs) and to other foreign affiliates. The results for Slovenia are on follows: positive horizontal spillovers from FIEs to domestic firms were found, and no significant difference from that result was found for spillovers from FIEs to other FIEs. The findings for Estonia regarding spillovers were just the opposite. The intra-industry spillover effect of FDI presence in Estonian manufacturing was insignificant for DEs in the same sector. These results stay basically the same for different specifications of the model. Our analysis also implies that there is no indication that other FIEs reap more possible benefits via FDI spillovers than DEs.

A policy implication of the analysis in this chapter is that providing incentives for FDI in general (or specifically for export oriented FDI) may be of dubious value in the FDI promotion strategies of the transition economies, at least as far as productivity is concerned. The existence of positive spillovers may depend on the level of economic development of the host country. Export oriented affiliates of MNEs are more than local market oriented FIEs interested in exploiting the host country's abundant production factors (see also, e.g., Kokko, Tansimi and Zejan 2001). For example, when these advantages have derived from relatively cheap labour rather than capital, then export oriented FIEs are not likely to have more positive effects on productivity of the host country than local market oriented FIEs.

Notes

1. For more references, see also the paper by Greenaway, Sousa and Wakelin (2004).
2. However, exceptions to this are, for example, the patent citations in patent applications by firms (see, e.g., the study by Globerman, Kokko and Sjoholin 1996).
3. FDI dummy $DUMF_{ijt}$ is equal to 1 if firm *i* (in sector *j* at time *t*) is and FIE, otherwise 0; $DUMEXP_{ijt}$ denotes the export orientation dummy, it takes the value of 1 if firm *i* exports at least 50% of its sales and 0 if it is more domestic market oriented.

4. However, note that in both these enterprise-level panel data-based samples, the share of FIEs is larger than in the aggregate sector-level data (see Table 5.5 for comparison).
5. There is a caveat in estimating the model as specified in this section, if the variable FDI_sector_{ijt}, instead of the definition used in this paper, were defined as simply the ratio of the sum of FIE assets to the sum of total assets of the sector. In that arguably inferior case, there might be difficulties in wholly separating the 'own-firm' and spillover effects from each other. This particularly would be a problem for the sectors with a small number of firms and one or a small number of FIEs making up large proportion of that sector, or in the case of one very large FIE entering the sector. It is therefore crucial to study and compare the estimation results also with the tables of descriptive statistics (tables 7.1 and 7.2) and employ the measure of FDI_sector_{ijt} where each FIE's own assets are subtracted from all the FIE assets of the sector. Naturally this sector-level FDI penetration variable now has different values for different firms, not only for different sectors. We have thus improved the results, by establishing a clearer difference between the 'own-firm' and spillover effects in the analysis.
6. The inverse of Mill's ratio is given by: $IMR = f(x)/(1-F(x))$, where $f(x)$ is the probability density function and $F(x)$ is the cumulative density function (Hardin, 1997).
7. $DUMF_{ijt}$ = FDI dummy variable (as a dependent variable); in the case of Estonia $DUMM_{ijt}$ = the majority FDI dummy variable; $PROD_{ijt}$ = Level of labour productivity; $EXPSALES_{ijt}$ = Share of exports in they sales of a firm; $LABC_{ijt}$ = Labour costs per employee of a firm; $FAPEREMP_{ijt}$ = Fixed assets per employee (measures the capital: labour ratio); $TFAPEREMP_{ijt}$ = Tangible fixed assets per employee (an alternative measure for the capital: labour ratio).

7
Results of a Fieldwork Project

Judit Hamar and Johannes Stephan

The second empirical analysis is based on a fieldwork project conducted between 2002 and 2003, which generated a large and unique database on 438 foreign subsidiaries in a selection of CEECs, namely the Czech Republic, Estonia, Hungary, Poland, Slovakia and Slovenia. The field work was done between 2002 and 2003 by the use of a concise, two-page questionnaire, sent out to the largest foreign investment subsidiaries in the countries named. The questionnaire is presented in the Appendix to this book (pp. 160–4).

In terms of methodology, the field work analysis focuses on the relationship between subsidiaries of MNEs which invested in CEE and their parent network, on the one hand, and the relationship between the subsidiaries and their local host economy, on the other.

Introduction

The fieldwork was designed to increase our understanding of the contribution of FDI to growth and productivity in the host countries. For our analysis of technology transfer via foreign direct investment in CEE, we departed from the usual methods in the respective literature:[1] in an extensive field work, we sent out an identical questionnaire to the top 50–100 foreign investors in Poland, the Slovak Republic and Hungary, and the top 20–50 in Estonia and Slovenia. A total number of 458 FIEs responded; their answers constitute our CEE subsidiary database. We analyse the data generated both at the firm level and at the aggregate level of subsidiaries belonging to our selection of countries. In this way, we work on an aggregated mezzo level while at the same time maintaining the advantages of having collected information at the firm level.

The objective of our analysis is to compare the potential of our selection of countries to benefit from technology and knowledge transfer. Conceptually, we focus on the determinants of internal and external technology and knowledge transfer. Derived from organisational theory and the international business and management strategy literature, we develop an innovative theoretical concept in the form of a four-quadrant taxonomy for FIEs. Potential for internal technology transfer is determined by the management relationship between parent and subsidiary and the ability of FIEs to adapt the foreign technology received from parents. External technology transfer potential is determined first by the potential for internal technology transfer and second by the role the host economy plays in the operations of the subsidiary.

The chapter starts by developing our theoretical concept of the taxonomy, which helps us to determine the potential for internal technology and knowledge transfer. Following a brief description of the main characteristics of the data we generated in our fieldwork, we located the positions of country-specific FIEs and sector-specific FIEs (according to technology intensity) in an empirical application of our taxonomy. The chapter then focuses on the potential for external technology and knowledge transfer to the host economy by assessing the role played by the host economy in the operations and management of the subsidiary. Here, the focus is on vertical, inter-industry links (backward and forward linking effects, OPT). This adheres to the general conclusion to be drawn from the literature that horizontal (or intra-industry) links between subsidiaries and the host economy produce either negligible or even negative technology effects. The chapter concludes the research using field work data by testing a model of the determinants of subsidiary development.

Conceptual framework: the taxonomy for FIEs and potential for technology transfer

The key empirical attempts to assess technology and knowledge transfer (hereafter, 'technology transfer') in the literature either use econometric methods at highly aggregate. Levels to measure directly the *extent* of the technology transfer (quantitative studies). Other analyses alternatively use deep-level interviewing techniques to identify the relevant *channels* of technology transfer. The quality of macroeconomic or industry-level studies hinges on the selection of proxies, in as much as proxies heed to be able to indicate technology and technological development via transfers from the foreign investor, and on the other hand heed to be available from a reliable statistical source. Those studies typically treat channels of technology transfer as 'black boxes', unable to shed light on the mechanisms within the channels. The advantage of qualitative studies is that they focus on uncovering the precise mechanisms within the channels of technology transfer, yet their results are typically not representative and hence lack generality. Their main insights remain within the case they have studied.

In our own analysis, we apply a method that combines qualitative and quantitative features: we use quantitative methods to analyse the data that we generated at the firm level. This data refer to qualitative matters of interest in as much as they are concerned with the management conditions existing within FDI inputs. The interpretation of how these conditions influence technology transfer is derived from organisational theory and the international business and management strategy literature.[2]

In this body of literature, the typical multinational investors are characterised as differentiated 'inter-organisational networks' (Roth and Morrison, 1992: 141) in which a variety of different FIEs often operate as 'quasi-firms' (Tavares, 2001). Within this network, each subsidiary is controlled through different mechanisms and to a different extent according to its role in the network. Hence, 'the subsidiary is a semiautonomous entity capable of making its own decisions but constrained in its action by the demand of head office managers and by the opportunities in the local environment' (Birkinshaw and Hood, 1999: 780).

In this characterisation, the intensity of technology transfer depends on its 'role in the network' and the conditions in the local environment: at the most general level, the literature assumes that the stronger the competencies of the subsidiary in terms of its own management *vis-à-vis* control by the head quarter, the stronger will be the positive technology-impact of the subsidiary on the host economy environment (e.g. Holm, Malmberg and Sölvell, 2002: 17, 29). Adhering to this approach, we assess three criteria pertaining to the network of relationships between the headquarters, the subsidiary and the host economy, and link these to the potential for technology transfer:

The role in the network

The 'role in the network' is conceptualised first by the management relationship between 'head office managers' and subsidiary managers: a dominant parent will manage the subsidiary on its behalf without much interference by the subsidiary's own management. At on the other extreme, an autonomous subsidiary is characterised by a mandate that lets it manage its own fate and the parent takes an inactive management role.

With respect to technology transfer, we can assume that the dominant parent will implement its foreign technology in the subsidiary, whereas an autonomous subsidiary will tend to take a more active role in the process. In particular, at early stages of the development of subsidiaries, parent companies can be 'adverse to technological incongruity' (Dyker and Stolberg, 2003, following Ozawa, 1979 and Wells, 1983) and can 'tend to place considerable stress on the importance of being able to impose their own technological culture on subsidiaries ... as a way of guaranteeing control over productivity' (Dyker and Stolberg, 2003: 4). Installation of an alien technology without the use of the adaptive expertise of the incumbent, however, pertains to a rather static process. The process ends with the installation of the parent's 'best practice' in the subsidiary, regardless of whether the technology functions efficiently in the particular host economy environment. The technology transfer process becomes dynamic with the subsidiary maturing and gradually assuming a more active role in the adaptation of the parent's technology. In a process of *technological interaction* between parent and subsidiary, the technological development of the subsidiary by way of technology transfer can be much more intense.[3] It the subsidiary matures in respect to its adaptive ability without a corresponding upgrading of its position in the management relationship (autonomy), however, the institutional learning curve will remain relatively flatter, as will the intensity of the technology transfer.[4] For our concept, we thus need the additional criterion of *adaptive ability*.

The role in the parent network

The role in the parent network is secondly characterised by the subsidiary's ability to adapt the parent's foreign technology to work efficiently in its own environment.[5] By enhancing its adaptive capabilities, the subsidiary establishes the process of technological interaction to the benefit of both partners, the parent and the subsidiary (Birkinshaw and Hood, 1999).

Those two criteria are used to assess the level of *internal technology transfer* – i.e. from parent to subsidiary.

External technology transfer

The third criterion pertains to *external technology transfer* – i.e. between the subsidiary and its host economy. It assesses the role that the host economy plays in the operations of the subsidiary. This is a straightforward concept, typically applied in the relevant literature.

In terms of methodology, those three criteria define the determinants of technology transfer which in turn act as necessary conditions for it to occur. Our indirect method allows us only to determine potential, not the intensity of the actual technology transfer.

With respect to the analysis of internal technology transfer, we simultaneously use the first two criteria and develop a two-dimensional taxonomy of FIEs to denote their transfer potentials (see Figure 7.1). On the vertical axis, we determine the FIE's position in the taxonomy according to its management relationship with the parent: FIEs operating under a dominant parent are located at the bottom half of the taxonomy. In terms of technology transfer, we assume that the potential for static effects is particularly high where the FIE has a dominant parent, willing and able to implement its own technology in the subsidiary. FIEs located at the top are more autonomous in the management of their own subsidiary.

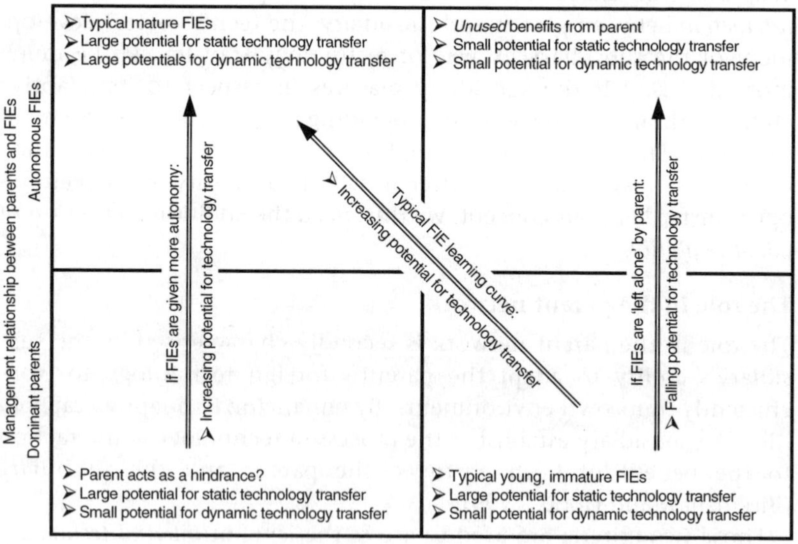

Figure 7.1 The conceptual taxonomy of FIEs and potential for technology transfer

Being autonomous, however, does not guarantee that the FIE management in fact reaps large benefits from its foreign investor: only if the subsidiary is able to adapt the foreign technology autonomously can technology transfer be intense and of a more dynamic type. The ability of FIEs to adapt the foreign technology they receive from their parents is depicted on the horizontal axis. FIEs located to the right have low adaptive abilities, whereas FIEs located to the left have high adaptive abilities.

In the graphical representation of the taxonomy, FIEs in the bottom-right quadrant feature typically young and immature subsidiaries, where adaptive ability is weak and the parent plays a dominant role in terms of managing the subsidiary. Whilse the potential for static technology transfer effects is large, the subsidiary receives the parent's technology (a dominant management position of the parent), and it is (so far) unable to contribute to its own technological development by adapting the foreign technology.

At on the other extreme, FIEs located in the top-left quadrant assume the highest position in terms of potential for both static and dynamic effects. Here, FIEs are not only more autonomous from their parent network in terms of management, they are also able to assume their own responsibility for the implementation and adaptation of this technology. Due to its high adaptive ability, the FIE will make use of the parent's technology, will be able to decide which technology to choose and how best to implement and adapt it (a static effect). When reporting back to the parent, a dynamic process of technology transfer between parent and subsidiary and back again can emerge. We assume that with FIEs maturing, they will typically move from the bottom-right to the top-left quadrant.

If an FIE is not granted additional autonomy in line with its increasing adaptive ability, however, then the parent will forgo the potential benefits from a dynamic interaction with its maturing subsidiary. Here, the subsidiary receives the parent's 'best practice', but is not allowed to participate by adapting it to function efficiently in its own environment despite its ability to do so. The potential for dynamic technology transfer remains low (bottom-right quadrant). FIEs located in the top-right quadrant of our taxonomy face the problem of not being able to use the foreign technology of the parent. Even if parent's technology is supplied to the subsidiary, and management is the responsibility of the subsidiary itself, the subsidiary's management is unable to implement and adapt the foreign technology it receives.

In the empirical part of the chapter, we use this theoretical taxonomy to estimate and compare the potential for static and dynamic technology transfer effects across the countries in our field study. We hence implicitly assume country differences between FIEs. Even if FIE differences within countries exist, or are more intense than country differences, our objective is to estimate any country-specific potential for technology transfer via FDI.

The main features of the sample

A concise two-page questionnaire was sent to FIEs in the five accession countries in 2002, and again in 2003 to improve the response rate and the representativeness of the country samples. Out of the 2,203 subsidiaries we approached with our questionnaire, only 458 provided us with a completed document. The response rate was the highest in Slovenia with 34.4 per cent, followed by Slovakia (30.2 per cent) and Estonia (30.0 per cent), while in Poland and Hungary only 18.8 per cent and 11 per cent respectively answered.[6]

The sample size seems to be quite small compared to the actual number of FIEs in the respective manufacturing industries (Tables 7.1 and 7.2). This is particularly true for Poland and Hungary. The 153 Polish FIEs and the 85 Hungarian FIEs, however, already provide a sufficiently large base to generate reliable and robust results. The lowest number of firms across countries and sectors are the 36 FIEs in the

Table 7.1 Distribution of firms, according to technology intensity[a] and Countriy

	High-tech		**Medium-high**		**Medium-low**		**Low-tech**		**Sum**[b]	
	(%)	*#*	*(%)*	*#*	*(%)*	*#*	*(%)*	*#*	*(%)*	*#*
Estonia	10.0	5	10.0	5	22.0	11	58.0	29	100.0	50
Hungary	11.8	10	22.4	19	25.9	22	40.0	34	100.0	85
Poland	9.2	14	36.6	56	34.0	52	20.3	31	100.0	153
Slovakia	4.8	3	27.4	17	30.6	19	37.1	23	100.0	62
Slovenia	5.8	4	39.1	27	40.6	28	14.5	10	100.0	69
Sum	**8.6**	**36**	**29.6**	**124**	**33.2**	**132**	**30.3**	**127**	**100.0**	**419**

Notes: *a* The classification scheme is derived from the usual OECD classification: high-tech (NACE 3530, 2423, 244, 30, 32, 33); medium-high tech (31, 34, 24 (excl. 2423), 352, 354, 359, 29); medium-low tech (23, 25, 26, 27, 28, 351); and low-tech (15, 16, 17, 18, 19, 20, 21, 22, 36, 37). See Hatzichronoglou (1997).

b Not all FIEs could be classified into the four technology classes, in particular service sector of the Estonian sample.

Source: Database from the fieldwork.

Table 7.2 Distribution of sample FIEs, by country and size

	Estonia	Poland	Slovakia	Hungary	Slovenia	All
No. of FIEs	73	153	78	85	69	**458**
Country share	15.9	33.4	17.0	18.6	15.1	**100.0**
Small FIEs [a]	23.4	24.2	34.2	14.1	38.9	**26.4**
Medium FIEs [b]	44.7	24.8	30.2	27.1	31.9	**29.6**
Large FIEs [c]	31.9	51.0	35.6	58.8	29.2	**44.0**

Notes: *a* Share of country-specific FIEs with up to fifty employees. *b* With up to 200 employees. *c* With over 200 employees.
Source: WP4 database from the fieldwork.

high-tech group. Again, this number appears to be sufficiently high to warrant statistical and econometric analysis. In terms of the distribution of FIEs across countries, Poland dominates the whole sample: the share of Polish FIEs is about one-third, which however corresponds to the comparatively large size of the Polish industry.

Comparing the sample distribution with the actual distribution in the sectors and countries, we identify some biases: the share of more sophisticated activities (high- and medium-high technology industries) is over-represented. This particularly pertains to our Polish sample. The sample distribution according to sales structures and export structures in each technology class is more similar to the actual structures in the countries and sectors. The Slovenian sample, however, includes more export oriented FIEs in the medium-high technology industries, while low-tech firms are under-represented. The opposite bias is true for the Hungarian sample, where high-tech FIEs accumulate only 14 per cent of the total exports of the country sample as compared to the 26 per cent share in manufacturing exports of all actual FIEs. Finally, our samples are biased towards larger FIEs. This is a common problem in large-scale fieldwork. The average size of firms in terms of employment is 450 employees across the whole sample. More specifically, the biggest shares of larger FIEs (with over 200 employees) are recorded in Hungary and Poland, the biggest share in small FIEs (with up to fifty employees) are in Slovenia and the Slovak Republic. Hungary records by far the lowest share in small FIEs.

By equity share, the sample has only a few minority foreign-owned FIEs (15 per cent), while 57 per cent of all FIEs are totally foreign-owned. The 100 per cent foreign-owned FIEs are the most frequent in the Estonian, Hungarian and Slovak groups (67 per cent, 64 per cent and 69 per cent, respectively), while in the Polish sample, 50 per and in Slovenia only 42 per cent were completely foreign-owned. By age of the sample FIEs, the Estonian and Hungarian FIEs were older than the average (69 per cent in both were established before 1995), while in the Polish, Slovak and Slovenian sample, more then half (54 per cent, 59 per cent, and 57 per cent, respectively) were registered as FIEs after 1995.

These differences by countries in the structure of sample FIEs partially explain the heterogeneity in the answers to some of the questions. This is why some links are statistically significant by countries, but not in the whole sample, or vice versa.

The empirical taxonomy of FIEs and potential for internal technology transfer

We shall now translate the data generated in our field work into proxies for our determinants of potential for internal technology transfer, analyse the data according to our concepts and locate our subsidiaries in an empirical application of our theoretical taxonomy. Because we are interested in the country-specific potential for technology transfer, we position country groups of subsidiaries rather than single subsidiaries into our taxonomy. With respect to the FIE maturity issue, we test whether we can detect significant differences between young and more mature FIEs in the relationship concerning FIE productivity growth and autonomy, and whether our assumptions about the direction of this relationship can be supported by our empirical data.

The tests are conducted by way of a Spearman-Rho rank correlations analysis between the individual indicators of autonomy and FIE productivity growth: the indicators from our field study are in discrete terms of four ranks and normalised in equal steps between 0 for complete autonomy and 1 for a most dominant parent.[7]

Autonomy in business functions and fields for initiative to change

In our fieldwork, we designed the questionnaire to provide us with the information needed to position our FIEs in our taxonomy. With respect to the management relationship, we asked firms who undertake FIE management in thirteen individual business functions and three distinct fields pertaining to taking initiative about changes in the subsidiary's operation. The business functions range from operational functions (including 'supplies and logistics', 'accounting and finance', 'operational management' and 'process engineering'), to market-related business functions (including 'market research', 'distribution and sales', 'after-sales services', 'advertisement' and 'marketing') and more strategic business functions which include 'product development', 'determining product price', decisions pertaining to 'investment and finance', as well as 'strategic management'. The three areas of initiative for change pertain to changes in the organisation of business functions, in the number of lines of business and in the composition of sales and exports.

Across the whole sample, our indicator for autonomy in business functions reaches an average of 0.25 for the group of operational business functions. The average for the group of market-related business functions signifies a more dominant position for the parent companies with an average level of 0.33 across all our firms. Average strategic business functions are typically decided upon mainly by the investor companies. In our panel, the autonomy indicator averages 0.44 in this group of functions. Our autonomy indicators for initiative for change are lowest for changes in the organisation of business functions (0.37) and highest for changes in the number of lines of business (0.47), indicating rather low autonomy in this field. Changes in the composition of sales and exports reach a medium level of 0.42.

Adaptive abilities are more difficult to assess in fieldwork by use of a questionnaire. Hence, we work with the plausible assumption that adaptive ability in a given group of subsidiaries is typically low if productivity growth since the advent of the foreign investor rises with a more dominant parent. That is: the more the parent is involved in the management of the subsidiary (and hence in the implementation of foreign technology in the subsidiary), the faster is productivity growth in the subsidiary. Likewise, adaptive ability is assumed to be rather high if productivity growth is particularly strong in subsidiaries with higher autonomy: here, FIEs which assume more responsibility for the implementation and adaptation of the foreign technology they receive from the parent are also the ones to experience the most intense productivity improvement. A positive and significant correlation between FIE productivity growth and the respective FIE autonomy indices (with 0 denoting autonomy) signifies low adaptive abilities, whereas a negative and significant rank correlation signifies rather more developed adaptive abilities. The tests are conducted by way of a Spearman-Rho rank correlations analysis because of the discrete nature of our data.[8]

The correlation analysis across the whole sample in fact establishes that most of our sample FIEs are able to increase productivity levels faster with rising dominance of the parent: the correlation coefficients are mostly positive but generally very low (between 0.09 and 0.15). Not all correlations, however, turn out to be significant: in the group of operational business functions, the relationship is typically negative yet statistically insignificant. It is significant and positive only for decisions taken on 'supplies and logistics'. Among the group of business functions targeted at market-related activities, all functions turn out significantly and positively correlated with FIE productivity growth. Among the group of more strategic business functions, all but 'investment and finance' produce significant positive correlation. In terms of the sources of initiative for change, our results across the whole sample also suggest that FIEs are able to increase productivity with an increasingly dominant position of the parent in the management relationship with the subsidiary: significant positive correlations can be established for initiating change in the 'number of lines of business' and in 'composition of sales and exports', but not however, in the 'organisation of business functions'.

These results suggest that across all our sample FIEs, adaptive abilities are generally rather low. Adaptive abilities are the most advanced in the more basic operational business functions and less so among the groups of more sophisticated functions and areas of initiative for change.

In sum and for the whole sample (Table 7.3), having a dominant parent prove to be one of the decisive factors in FIE development: the more dominant are parents, the faster and the more stringently are efficiency-improving changes implemented in the FIE, and the more the parent company network assumes responsibility for the more sophisticated business functions, the deeper is the reform process at

Table 7.3 Autonomy indices[a] and FIE productivity growth in the complete sample

	Operational	**Market oriented**	**Strategic**	**All**
Autonomy indices	0.25	0.33	0.44	**0.34**
Correlation coefficient[b]	0.012	0.138**	0.111*	**0.119***
Significance	0.800	0.004	0.020	**0.013**
N	441	438	441	**441**

Notes: *a* The autonomy indices are averages over the three groups of business functions, and over the total sample in the last row.
b Correlation significant at the 0.05 level (2-tailed) are denoted with a single *, correlation significant at the 0.01 level (2-tailed) with **.
Source: WP4 database from fieldwork.

the receiving end. This turns out to be particularly true for market-related activities and for the more strategic functions, but less so for the more basic operational functions.

Tests of the maturity assumption

In terms of the maturity issue, we can hence conclude that our sample FIEs are already developed enough to have achieved some autonomy over operational business functions, but are still not sufficiently developed to assume a more decisive management role in the more sophisticated functions that could accelerate their technological development and contribute to the technological development of the MNE network (compare this interpretation with, e.g., Moran and Bergsten, 1998 and, for 'open networking' or 'strategic technology transfer', Dyker and von Tunzelmann, 2001). This in fact is not a surprising result at all: the typical learning curve for FIE subsidiaries with respect to their autonomy over business functions would usually start from operational autonomy in marketing and finally strategic autonomy (see Majcen, Radošević, and Rojec 2003c: 12–13). In an additional control for the age of FIEs, we establish a weak correlation between the number of years since registration as a FIE and the average over all business function autonomy indices with the expected negative sign. This result pertains both to the business functions and to the initiative for operational and strategic change. In terms of a cross-country comparison, we can establish a positive relationship between the level of development of the host economy and the autonomy of its FIEs. This is particularly pronounced for the operational and more strategic business functions.

An analysis of rank correlations accounting for the age of FIEs in fact suggests that the FIEs which were established more than ten years earlier appear to depend less on an active or even dominant parent with respect to their own productivity development than those established less than five years earlier.

Our results reaffirm the 'usual learning curve of the FIE', and we deduce from this that FIEs in CEECs are still rather low on their institutional learning curve – i.e. having achieved some degree of autonomy only in operational functions. This is particularly pronounced among our Slovak FIEs, and least pronounced among the Polish firms: here, the degree of autonomy in market-related business functions is regularly higher than for operational or more strategic functions, which clearly distinguishes Roland's FIEs from all other FIEs in our other country panels. The large size of the Polish market, with the resulting bias on a more market (and hence domestic) orientation of the Polish FIEs may account for this difference. In fact, across all our samples, the more domestically oriented FIEs are, the more autonomy they tend to have (a significant and positive correlations with most business functions).

The location of sector groups of FIEs in the empirical taxonomy

On the horizontal axis, we determine the FIE group's position in the taxonomy according to the strength and the sign of the correlation between FIE productivity growth and individual levels of autonomy in business functions and areas of initiative for change. This location on the axis tells us whether the group of FIEs assessed benefited more from a dominant parent (a positive correlation) or rather more from a more active role in the management of their own subsidiaries (a negative correlation). Knowing whether FIEs are able to increase productivity with either high or low autonomy, however, does not yet tell us whether the FIEs are in fact fully autonomous or rather les so. Hence, we determine secondly the FIE group's position in the taxonomy according to its actual autonomy.

We depict the sector averages of the autonomy indices for business functions and initiatives for change individually on the vertical axis. Obviously, because our taxonomy is a two-dimensional one, we present business functions only where we can establish significant correlations with FIE productivity growth. Lacking benchmarks of typical autonomy levels, we normalise our autonomy levels around the median levels for each category individually. This way, we can locate each business function and area of initiative for change vertically around what we perceive as the 'normal' level and interpret the levels as above and below average.

This graphical representation of the taxonomy has the advantage that we are able to determine graphically the FIEs' potential for static and dynamic technology transfer. The disadvantage of this conceptual taxonomy is that the vertical axis measures a criterion included into the criterion on the horizontal axis. This, however, is owing to the limitations of data collectible from firms in fieldwork by use of a questionnaire.

The first grouping of our sample focuses upon a classification of FIEs according to their belonging to typical technology intensity classes (the OECD classification, see Hatzichronoglou 1997).[9] Initially, we would assume that the higher the technology intensity, the more intense will also be the control of the FIE by a dominant parent: the parent network can benefit most if the FIE adheres to high-quality and precision standards in industries such as pharmaceuticals, medical, precision and optical instruments manufacturing. In technologically less sophisticated industries such as the food, textile, or wood processing industries, close control of MNCs' standards may be of less relevance.[10]

In fact, the class most intensively integrated into their parent networks appears to be the high-technology group, with an overall indicator of slightly more than 0.37. In comparison to the total sample's average, this is particularly pronounced for the operational and strategic type of functions and less for the market-related business functions. This result is supported by the fact that the equity share of parents in the FIEs of this sub-panel is higher than in the average over all the FIEs. Despite their relatively low competency mandate, our high-tech FIEs have benefited relatively little in terms of productivity, technology or quality improvements: most correlations turned out to be insignificant. We can establish a strong and significant positive correlation – among market-related functions – only for market research with a coefficient of 37 per cent, and a coefficient – among everyday operations – for operational management, which turned out to be significant but negative and even higher, at 0.45 per cent.

Figure 7.2 depicts the location of the sector-specific groups of FIEs in our taxonomy: the high-technological FIEs assume a location towards the lower two quadrants of Figure 7.2, with operational functions on the far left and market-related functions on the far right. In terms of our concept and the two criteria for technology transfer, we would hence conclude that, with respect to operational functions, the potential for static technology transfer is large, while that for the dynamic technological interaction between parent and subsidiary is smaller. FIEs operate under a dominant parent and are at the same time successful in adapting the foreign technology they receive from the parent to work efficiently in the environment of their own host economy. Alas, the FIEs involved are not sufficiently autonomous actually to make full use of their adapting ability on a large scale. The same result with respect to static and dynamic potential appears to apply to market oriented business functions: there is a large potential for static technology transfer through intense headquarter control, yet FIEs are less successful in the adaptation of foreign technology.

With FIEs maturing along the typical learning curve, we would expect increasing potential for dynamic technology transfer between parent and subsidiary. Further FIE development thus crucially depends on whether those FIEs are in fact given more autonomy. If they are,

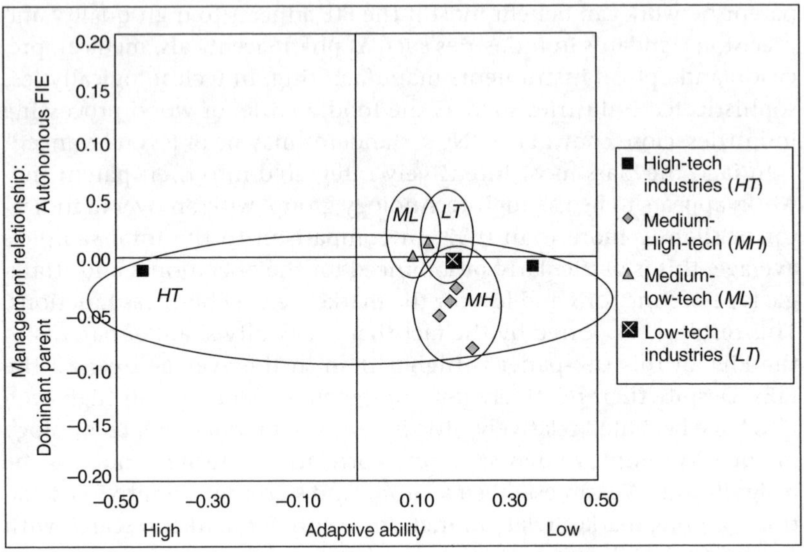

Figure 7.2 Empirical taxonomy of FIEs and potential for technology transfer: sector-specific groups of subsidiaries

Note: The points marked in the taxonomy correspond to individual business functions and areas of initiative for change. We have rescaled the vertical axis to present the FIEs' location in their autonomy relative to the average of the total sample. The location of ellipses around the points of groups of FIEs is determined vertically by the largest and lowest values for the autonomy indicators of all business functions and initiatives, whether significant in the correlations or not.

however, (an OTP-type of business), then little progress can be expected. Surprisingly, however, within our high-tech group very few FIEs can be considered to be an OPT-type.

The most autonomous technology classes turned out to be the medium- low-technology industries with an average indicator value of slightly below 0.33, and the low-technology industries with a value of slightly above 0.33. In both groups, we can establish significant positive correlations for only very few business functions with weaker coefficients between 12 and 20 per cent. With medium- low-technology FIEs able to develop faster under the guidance of a dominant parent, yet already being quite autonomous, we are led to conclude that there must be some unused benefits in terms of technology transfer (Figure 7.2, upper-right quadrant). Ther potential for both static and dynamic technology transfer hence appears to be rather small: the intensity of headquarter control by the foreign technology-bearing parent MNC network is low despite the fact that the FIEs' level of maturity (a positive correlation) still necessitates an active parent to fully benefit from technology transfer potential. Considering additionally that the low-wage comparative advantage of the region can be assumed particularly important for FIEs in industries with a rather lower technological sophistication, this interpretation of our concept appears to be well founded. Whether or not those FIEs mature in terms of their ability to learn to adapt the foreign technology they receive cannot be answered within this framework.

In the case of medium- high-technology FIEs, we can establish significant positive correlations for all market-related, some strategic business functions and two out of three sources of initiative for change. The correlation coefficients in these business functions and sources of initiatives are in fact higher than for the average over all FIEs. With an overall autonomy indicator demonstrating a rather low competency mandate within the parent companies' networks (slightly below 0.37), we are here presented with a typical picture of immature FIEs low on their learning curve (Figure 7.2, lower-right quadrant). In terms of our concept, this suggest that significant potential for static technology transfer is already prevalent while the adaptive abilities of these firms is still rather low, suggesting a lower potential for dynamic technological interaction between the parent and the subsidiary. However, with these FIEs maturing on the typical learning curve, we can assume increasing potential.

The location of country groups of FIEs in the empirical taxonomy

In a cross-country comparison of autonomy, our Slovenian FIEs appear to be the most autonomous, with an overall indicator of 0.30. This is particularly pronounced for operational functions but is also true for strategic business ones. Only in the group of market-related functions is autonomy below average, and the sources of initiative for change in Slovenia suggest more autonomy than across all countries, exceeded only by our Estonian FIEs. Our Slovenian FIEs nevertheless exhibit a strong relationship between autonomy and FIE development (particularly in market oriented functions, strategic business functions and sources of initiative), with correlation coefficients of between 25 and 39 per cent. Apparently, our Slovenian FIEs are well placed to benefit from their cooperation with their parents (high correlation coefficients), only they are already considerably autonomous. We would therefore locate our Slovenian subsidiaries in the upper-right quadrant of our taxonomy (Figure 7.2). According to our concept, we can thus assume some unused benefits from the relationship and level of interaction with the parent; the potential for technology transfer, whether static or dynamic, is rather low. We cannot, however, deducet from this interpretation whether those FIEs will in fact mature to move into the top-left quadrant (thereby learning to adapt the foreign technology to the particularities of the host economy) or rather remain stuck in their current position.

At the other extreme, the Slovak Republic's FIEs seem to be the least autonomous, with an average indicator over all business functions of 0.42. Here, particularly the market-related business functions and the strategic functions assume much lower autonomy as compared to the other countries. In terms of sources of initiatives for change, autonomy is by far the lowest among the country groups. Some of this may be attributable to the fact that in the Slovak Republic, large-scale FDI is of much later origin (an average age of 7.8 years since their registration as FIEs as against 8.8 years for the whole sample and 9.1–10.2 years for Estonia and Hungary). Foreign investment also until recently involved a comparably higher extent of political uncertainty, suggesting more intense control by the parent companies. However, only market-related business functions and the initiative for change in sales and exports are significantly and positively correlated with FIE development, the correlations for operational and strategic business functions are insignificant. This assigns our Slovak FIEs a position in the lower-right quadrant of our taxonomy. Apparently, Slovak FIEs are particularly well placed to benefit from an active parent role in FIE management in market-related fields, suggesting a rather low ability to adapt the foreign technology received. According to our concept, this would suggest a large potential for static technology transfer, but a rather small potential for the dynamic effects of technological interaction between subsidiary and parent.

While these results correspond to those of the low- and medium-low-tech industries, the Slovenian economy has a particularly high share of medium-high technology FIEs (39 per cent, against 27 per cent for the whole sample). Hence, our assessment allows us to assume a rather more optimistic view on the potential for this class of Slovenian FIEs. If we assume that our future Slovak FIEs mature along a typical FIE learning curve, then we can expect a rising potential for dynamic technology transfer effects.

Our Hungarian FIEs appear to be higher up the institutional learning curve with above-average autonomy in a number of business functions,

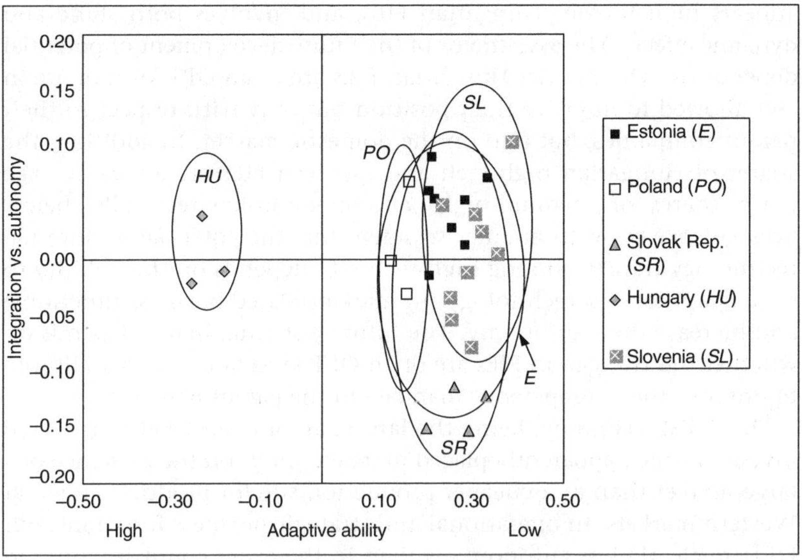

Figure 7.3 Empirical taxonomy of FIEs and potential for technology transfer: country-specific groups of subsidiaries

Note: The points marked in the taxonomy correspond to individual business functions and areas of initiative for change. We have rescaled the vertical axis to present the FIEs' location in their autonomy relative to the average of the total sample. The location of ellipses around the points of groups of FIEs is determined vertically by the largest and lowest values for the autonomy indicators of all business functions and initiatives, whether significant in the correlations or not.

mainly strategic and operational, but less in market-oriented functions (the overall autonomy indicator is 0.33). However, not many significant correlations between business functions and sources of initiative for change with FIE development could be found: among those business functions significantly correlated with FIE development, all show negative signs, implying that our Hungarian FIEs on average are able to adapt and implement some of the foreign parent technology to the particularities of the host economy environment under their own responsibility (i.e. upper-left quadrant). With Hungary being the country in the region with the longest history of large-scale FDI, this result is not surprising and lends further support to our conceptual framework: our Hungarian FIEs are on average more mature than our FIEs in the other countries. The potential for technology transfer appears high for our Hungarian FIEs, and involves both static and dynamic effects. The assessment of the future development of potential depends on whether the Hungarian FIEs are of an OPT kind or are in fact allowed to improve their position not only with respect to their parent companies, but also on the domestic market. In addition, the shares of Hungarian high-tech and low-tech FIEs are above average (with shares of medium-high- and medium-low-tech FIEs below average). We know from our assessment that the potential for internal technology transfer among high-tech FIEs depends on their ability to both adapt foreign technology (in market-related business functions) and increase their autonomy. The future potential hence depends on whether the Hungarian FIEs are of an OPT kind or are in fact allowed to improve their competency mandate in the parent network.

The Polish economy, being the largest among our CEECs, attracted investors which apparently placed more emphasis on the existence of a large market than as a cheaper production site for products aimed at Western markets. In operational and strategic business functions, our FIEs are clearly less autonomous than in the other countries; only in the market-related functions do our Polish FIEs assume more of their own responsibility and be able to initiate change (to sales and exports). Additionally, few significant relationships can be found between autonomy and FIE development: significant positive correlations occur among market-related and more strategic business functions, but with coefficients as low as below 9 per cent. In our taxonomy, our Polish FIEs would thus be located in the right-hand two quadrants, with market-related business functions tending to the upper-right quadrant and strategic functions to the bottom-right quadrant. Hence, we can tentatively conclude that our Polish FIEs lave until now experienced little potential for technology transfer in market-related functions, yet a larger potential for static technology transfer in the more strategic functions.

Our Estonian FIEs also assume a middle rank in terms of average autonomy, with an overall level of 0.32. In market-related functions, however, autonomy is comparatively high and surpassed only by Poland; in the sources of initiative for change, autonomy is the highest amongst all country groupings. In contrast to the Hungarian and Polish FIEs, our analysis establishes a large number of positive and significant correlations, with coefficients above 20 per cent. Most positive and significant correlations appear in market-oriented business functions. In total, we can assign our Estonian FIEs to the upper-right quadrant of our taxonomy. In the empirical application of our taxonomy, the long tail down to the bottom-right quadrant originates from only one business function (investment and finance), exhibiting extremely autonomy way below the average of the other functions. According to our concept, we can conclude that our Estonian FIEs can theoretically benefit from their parent, yet are too autonomous to experience any large potential for technology transfer. This result corresponds to the fact that the Estonian economy accommodates by far the largest share of low-tech and the second-to-highest share of high-tech FIEs.

Intensity of integration with the host economy and potential for external technology transfer

In the following part of our analysis, we assess the potential for foreign technology to diffuse from the subsidiary to the host economy. We assume that two *sine qua non* conditions have to be met simultaneously for external transfer effects to materialise to an economically significant extent: first, a high potential for internal technology transfer is required. It this does not occur, there is no technology that can actually diffuse to other firms in the host economy. Second, for intense technology to transfer to the host economy, the subsidiary has to be intensively integrated with other firms in the host economy. For our analysis, we focus on vertical links only and do not distinguish between technology transferred that is internalised by contracts or pure spillovers. The horizontal technology transfer between firms of the same industry is not covered, because it is very difficult to assess in our mezzo-method and because we draw the conclusion from other empirical research that horizontal effects are after insignificant and sometimes even negative. In terms of analysing potential, this channel of technology transfer thus provides little insight.

In particular, the size of the potential for vertical technology transfer is tested by assessing the share of host economy firms in the subsidiary's procurement, giving rise to a potential for backward linking effects; we also assess the share of the host economy in sales (domestic buyers) to test for any potential for forward linking effects. The analysis of backward linking effects not only focuses on the share of local suppliers in terms of material and prefabricated products, but additionally on the supply of services to the subsidiary. These include in particular sources of finance (assuming that host economy banks may also provide a channel for technology transfer), assistance in quality control (which in CEECs is often thought of as assistance in

International Standards Organisation (ISO) total quality control certification), the supply of patents, licences and R&D results and ideas (which in FIEs will typically originate from the parent's network), the supply of qualified workers and their training, and finally the supply of qualified managers (the latter two are typically determinants of the quality of the location).

Destination of sales and origins of procurement: material linkages

Across the whole sample, our FIEs export slightly more than half (52 per cent) of their sales, which include sales to a foreign parent as well as sales to other foreign buyers (see Table 7.4). The most export oriented FIEs are in the Slovenian (73 per cent) and in the Slovakian samples (64 per cent). Our Hungarian and Estonian sample FIEs each export only 52 per cent of their sales, while Polish FIEs are much more local market oriented (33 per cent). Within exports, direct sales to the parent firms in the sample average are 30 per cent, with significant differences between country samples: the lowest rate is recorded for our Polish FIEs (21 per cent), next comes Hungary (28 per cent) and Estonia (24 per cent), while the Slovenian and especially the Slovak FIEs export their products to a much greater extent directly to the parent firm (38 and 48 per cent, respectively).

Sales directed to the host economy can take two forms: first, FIEs can sell to the host economy domestic market, and second, FIEs may also sell to other domestic subsidiaries within their own parent investor's network. While the latter category assumes rather low shares, the

Table 7.4 Distribution of sales and purchases, by regions, average percentage shares of sales/purchases by country-specific FIEs

	Estonia	Poland	Slovak Rep.	Hungary	Slovenia
Sales to:					
Foreign parent	23.8	20.8	47.5	27.7	37.8
Other foreign buyers	28.4	12.0	16.9	24.4	34.9
Domestic network FIEs	3.5	4.5	2.5	3.5	0.5
Other domestic buyers	45.7	62.6	31.7	43.3	28.3
Purchases from:					
Foreign parent	23.8	34.0	36.0	17.9	23.0
Other foreign suppliers	29.7	17.8	23.0	32.0	34.8
Domestic network FIEs	4.5	6.7	1.6	1.2	0.5
Other domestic suppliers	38.7	40.5	37.6	45.3	41.6

Source: WP4 database from the fieldwork.

shares of sales to other domestic buyers range between 63 per cent for our Polish FIEs (which reflects their considerable domestic market orientation) to only 28 per cent across our Slovenian firms. The Estonian share of 46 per cent appears quite high considering the small size of the domestic market, but may be attributable to the large share of financial services, which are naturally directed to the host economy. Slovakia's and Hungary's shares of 38 and 45 per cent, respectively, reflect the respective sizes of their domestic market.

Concerning the regional distribution of procurement, we find shares of one-third for supplies from parent networks (where the parent itself is the dominant supplier, other network partners reach only negligible shares), other foreign suppliers and other domestic suppliers across all sample FIEs. Again, large differences between country groups exist. In this respect, our Polish and our Slovak FIEs stand out because they procure much higher shares of purchases from the in parents (around 35 per cent). The shares of the other countries are more than 10 percentage points lower, with Hungary the lowest (18 per cent). Our Polish and Slovak FIEs also procure to a much lower extent from other foreign suppliers: here shares are again some 10 percentage points lower when compared to the other three country's FIEs. Other domestic suppliers reach comparable levels across all country-specific averages (some 40 per cent).

In terms of potential for vertical, external technology transfer derived from the analysis of distribution of purchases and sales, we assume that potential is particularly high where subsidiaries purchase large shares of their supplies from the host economy and at the same time sell large shares to their parent network. This plausible assumption is based on the expectation that subsidiaries in this constellation will particularly demand a high technological level (in terms of quality of produce) and precision (in terms of timely delivery of supplies) (backward linkages); this will be the more intense the more the subsidiary sells to their parent networks. A higher potential will also pertain to subsidiaries purchasing larger shares from their parent networks and at the same time selling large shares to the host economy. This reflects the usual assumption that intense integration with the host economy increases the potential for technology transferred from the parent to the subsidiary to actually find its way to the host economy (forward linkages). The test of these linkages, however, has to be conducted at the firm level.

From the data of our fieldwork, we identify the largest potential for purchase-driven, backward-linking technology transfer in the case of

Table 7.5 Share of backward and forward linking FIEs and the OPT type of FIEs, percentage shares of FIEs fulfilling the criterion (all FIEs in country groups = 100)

	Estonia	Poland	Slovak Rep.	Hungary	Slovenia
Backward linkages					
50% criterion	8.2	7.8	12.8	4.7	13
80% criterion	4.1	0.7	9.0	1.2	0.0
100% criterion	0.0	0.7	0.0	0.0	0.0
Forward linkages					
50% criterion	12.3	26.1	14.1	3.5	2.9
80% criterion	6.9	13.7	9.0	0.0	0.0
100% criterion	1.4	6.5	3.9	0.0	0.0
OPT					
50% criterion	8.2	11.1	23.1	10.6	13.0
80% criterion	2.7	4.6	10.3	5.9	8.7
100% criterion	1.4	0.7	3.9	2.4	0.0

Notes: The three criteria pertain to the share of purchases or sales directed to the domestic market or the parent network.
Source: WP4 database from the fieldwork.

the Slovak Republic, and to a lesser extent also in Estonia and Slovenia. In the latter, however, we cannot identify any FIEs that purchase more than 50 per cent from the host economy and sell more than 50 per cent to their parent network. The countries in which we expect the weakest backward linkages via procurement of material and semi-finished products are Hungary and possibly Poland.

With regard to sales-driven forward linkages, the Slovak Republic and Estonia again appear to have comparably high shares of FIEs, fulfilling our criteria on of purchasing from parents and selling to the host economy. Here, however, Poland exhibits the highest shares, which is not too surprising in new of the country's large domestic market and the subsequent local market orientation of Polish FIEs. Hungary again ranks at the bottom of the list, and this time the assessment for Slovenia unambiguously suggests a rather low potential for technology transfer of a forward linking kind.

The analysis of purchasing and selling structures also provides us with some indication on the whether OPT type of FIEs are a widespread phenomenon in our country samples: such FIEs are characterised mainly by buying from their parent networks and at the same time mainly selling to their parent networks. In this respect, we do find a significant positive correlation between selling to the foreign partner network and procuring from the parent network across all our FIEs, albeit with a low coefficient. For the identification of the OPT type of FIEs, we additionally compare the shares of FIEs in total country-specific FIEs that largely sell and at the same time largely purchase from their parent networks. The results of this analysis establish that the Slovak panel clearly contains the largest shares of the OPT type of FIEs, followed at a significant distance by Hungary, Poland and Slovenia. The lowest share of FIEs of the OPT type is to be found in our Estonian panel.

Channels for non-material external technology transfer

Two further channels for backward linkages of a more non-material type can be assessed from our data: the role the host economy plays for supplying management areas that may be important for the competitiveness of subsidiaries, and the role domestic money and capital markets play as sources of FIE finance. In terms of areas of competitiveness (Table 7.6), we assume that the large potential for such vertical, non-material and external technology transfer exists where FIEs are being supplied to a large extent by the domestic economy in those areas of competitiveness the subsidiary itself determines to be particularly important. With respect to sources of FIE finance, we follow the usual and most straightforward method of comparing the weights of domestic, foreign and FIE own-resources.

Among the four areas of FIEs' competitiveness (Table 7.6), all turn out to be important on average across the whole sample. The most important ones turn out to be 'quality control assistance' and 'management'. The quality and supply of 'people and training' for personnel prove to be of slightly less importance, while 'patents, licences and R&D' turn out to be the least important. Across the country sub-samples, the results are in fact very similar: Hungary stands out because quality control is much more important than in the average sample and patents, licences, and R&D much less. In Estonia, people and training proves to be much more important than for the other countries.

The acknowledgement of these three important fields as areas of competitiveness also coincides with FIE development: across the whole sample, we can establish significant and positive correlations of the magnitude of 8–20 per cent between the level of importance of productivity growth, the improvements in the level of technology of production equipment and the level of quality of produce. The correlations are much higher for Hungary and Slovenia; for the Slovak Republic, we could establish only one significant correlation, and for Estonia none at all.

Table 7.6 Indicators for non-material external technology transfer potential

	Estonia	Poland	Slovak Rep.	Hungary	Slovenia
Areas of competitiveness[a]					
Quality control assistance	16.4	37.9	18.0	15.3	36.2
Patents, licences, R&D	4.1	23.5	3.9	3.5	11.6
People and training	5.5	23.5	10.3	1.2	11.6
Management	12.3	25.5	12.8	5.9	17.4
Average of the four areas	9.6	27.6	11.3	6.5	19.2
Sources of finance from[b]					
Retained earnings	0.71	0.61	0.80	0.73	0.70
Foreign parent	0.65	0.67	0.63	0.53	0.57
Other foreign sources	0.29	0.32	0.25	0.17	0.28
Domestic network FIEs	0.03	0.22	0.09	0.05	0.04
Other domestic sources	0.50	0.48	0.39	0.63	0.46

Notes: *a* Share of firms fulfilling the criterion. The criteria are defined as FIEs both considering the respective area as particularly important for their competitiveness (indicator ≥ 0.5), *and* valuing local sources (without the FIE itself) to supply those areas as equally particularly important (average indicator for local sources ≥ 0.5).

b The average country level of importance of each source, whereby we translat the answers into: 0 = Not important; 0.25 = Not very important; 0.5 = Important; 0.75 = Very important; 1 = Extremely important.

Source: WP4 database from the fieldwork.

For our analysis of the role of the host economy in FIE development in general and the potential for vertical technology transfer in particular, we asked firms about the sources supplying these areas of competitiveness. Those sources range from the own-subsidiary, domestic and external markets, the typical industrial and research networks, to the parent company network. According to our straightforward assumption, the potential for technology transfer rises with the intensity with which FIEs use local sources for the supply of areas of competitiveness they consider themselves as being particularly important. Institutions in the host economy constituting sources for such areas of FIE competitiveness do not only grant the subsidiary a link to the host economy. Technological development in the host economy is also induced: by supplying quality control assistance to the FIE, the local institution will learn to assess new technology and hence accumulate knowledge it can subsequently use in the other local firms it supplies; in supplying patents and licences via own-R&D, we can assume that the supplying company or research institution cooperates closely with the FIE to match demand; the supply of people, training – and in particular management – pertains to the technological development of human capital in the host economy.

Counting the number of firms that value an area of competitiveness as particularly important and value local sources to supply this area as equally particularly important, we can compare the roles of host economies across our country samples: the share of FIEs fulfilling these criteria are largest for our Polish FIEs in all four areas of competitiveness; the average share amounts to nearly 28 per cent. The second largest average share is recorded for our Slovenian FIEs (19.2 per cent). The shares of the group of Slovak and Estonian FIEs are around 10 per cent, and that of our Hungarian group is the lowest (some 6.5 per cent). In all country samples, FIEs are most intensively networked with the host economy in the area of quality control assistance, with a considerable gap in the area of management and people and training. In Hungarian and Estonian FIEs, the area of people and training shows comparatively weak links to the host economy, despite the above-average importance attached to this area of FIE competitiveness by the latter country's FIEs.

From the point of view of intensity of business networking of FIEs with their host economy, our fieldwork results therefore suggest that the Polish FIEs probably contain by far the largest potential for vertical technology transfer from this source, followed, with a significant lag by Slovenia. The Hungarian economy probably benefits the least from its FIEs in this respect, and the Slovak Republic and Estonia are somewhere in the middle. In all samples, the largest benefits are set to arise from linkages with local institutions providing quality control assistance to foreign direct investors' subsidiaries.

Finally, the potential for non-material, vertical technology transfer may also depend on the involvement of local sources of finance such as banks, domestic investors, the capital markets, and other domestic subsidiaries of the parent network. In our fieldwork, we again assessed the importance of a set of different sources: in general, our FIEs' main sources of finance turn out to be their own retained earnings, followed by the foreign investor parent. Only in the cases of Poland was the foreign investor parent slightly more important than retained earnings. Additionally, our Polish FIEs also procure finance to a significant extent from other domestic subsidiaries of the parent investor's network. In Hungary, other domestic sources (probably mainly banks) turn out to be second after retained earnings. This may be a reflection on the fact that foreign direct investors in some cases engage in Hungary in cooperation with foreign banks. Not surprisingly, the level of importance that attaches to retained earnings increases with the age of the FIE (again tested for the group of FIEs younger than five and older than ten years over the complete sample).

Comparing the levels of importance attached to domestic and foreign sources, the latter (including the foreign owner company and other foreign sources such as domestic banks, local investors, etc.) turn out to be more important than the former. This is particularly pronounced in our Estonian and Slovak firms, whereas for our Hungarian FIEs, domestic and foreign sources are almost equally important. The highest level of importance of domestic sources attaches to our Polish and Hungarian subsidiaries, followed, with some lag, by Estonian, Slovenian and Slovak FIEs. In terms of the role the host economy plays for the FIEs in providing sources of finance as an indicator for the potential for technology transfer from the subsidiary to the host economy, we can conclude that the potential from this channel is probably highest in Poland and Hungary, and much lower in the other countries.

The potential for internal and external technology transfer: summary of results

The analysis of the potential for vertical technology transfer complements our results generated from the analysis of the location of country-specific FIEs in the taxonomy. The taxonomy would suggest that our Hungarian FIEs contain quite a large potential for internal technology transfer and display a relatively intense adaptation of foreign technology received from their parents. Our Hungarian FIEs are thus well endowed with the conditions for an intense internal dynamic technology transfer between parent and subsidiary. In our analysis of external technology transfer potential, however, we establish that both material and non-material vertical links to the host economy suggest a rather limited potential for external technology transfer. Only with respect to the sources for finance does our analysis suggest an intense role for the host economy in the operations of foreign investment subsidiaries in Hungary. This could be interpreted to signify what is typically termed a 'dual economy': well developed and mature subsidiaries but with little contact to the host economy.[11] Additionally, the share of the OPT kind of FIEs appears to be significant in Hungary.

Our Estonian and Polish FIEs play a comparatively important role in their host economy, in both forward linking business and (in Poland's

FIEs) also in supplying areas of competitiveness and serving as sources of finance. In the taxonomy, however, both countryies, FIEs featured a 'premature autonomy' and an inability to adapt the foreign technology to their own needs. In the case of Poland, this is mainly due to the strong market orientation of FIEs. The potential for external technology transfer would thus be significant in both countries, if only our FIEs signalled a larger potential for internal technology transfer: in their current situation, the potential for technology transfer via FDI subsidiaries is rather low for both countries.

In the case of our Slovenian FIEs, the analysis suggests a rather low potential for internal technology transfer, mainly rooted in the lack in adaptive abilities. At the same time, vertical linkages with the host economy for sales and procurement are comparatively less intense. Only with respect to non-material linkages supplying areas of competitiveness and FIE finance can we establish an above-average role for the country's respective host economies. In total, however, our analysis suggests a rather limited potential for technology and knowledge to diffuse from parent to subsidiary and further on to the host economy.

In the case of our Slovak FIEs, the potential for technology transfer today appears low according to our taxonomy, yet with FIEs maturing a brighter future may lie ahead. In particular, the conditions for intense dynamic technology transfer between parent and subsidiary are well in place and await their exploitation. With regard to the conditions for a high potential for external technology transfer, the results were rather mixed: in our analysis of backward and forward linking, we established intense networking activities, but also a high share of FIEs fulfilling our criteria for an OPT kind of subsidiary. The intensity of non-material linkages is likewise only average across our country samples. In sum, we have to conclude that there is a rather small potential at this point of time, but we can expect this potential to increase in the future.

The determinants of FIE development

So far, our case study analysis of technology transfer has focused on a comparative analysis of the potential for technology transfer between the countries of our sample. While this analysis has assessed direct and indirect transfer potential separately by looking at the autonomy/adaptive ability issues for technology transfer between parent and subsidiary and the integration with the host economy for external technology transfer, respectively, the following analysis amalgamates all data in one cross-sectional regression analysis. What factors determine the development of subsidiaries? How do these factors interrelate? This analysis can be viewed as supplementary to our earlier analysis of the database.

Our fieldwork focuses on a set of conditions for swift FIE development, and a number of indicators for FIE technological development itself. This input–output design allows us to use fairly simple methods for evaluation of the conditions for FIE development, or for the intensity of technology transferred from the parent company to the FIE. The indicators include a set of variables indicating the extent of FIE autonomy from the parent company, a set of variables classifying the mandate or purpose of engagement of the parent companies in CEE firms and other subsidiary-specific determinants. Positive FIE technological development is measured in terms of increases in productivity in production, in technology of production equipment and in the quality of produce.[12]

To produce a clearer picture of the relationship between all the individual indices and FIE development, as well as their interdependence (in terms of relative importance), we apply a simple log-linear LS regression analysis. This additionally includes independent variables not assessed above, such as the size of the subsidiary (in terms of numbers employed) and the product scope of the subsidiary (measured by the number of lines of business per employment). We also control for the influence of final product-producing FIEs, for greenfield investments and for majority-owned subsidiaries on FIE development. As dependent variables, we use our average FIE development index, as above. All data in the regression are the form of natural logarithms.

Our regression formula reads:

$$\pi = \beta_1 OP + \beta_2 MA + \beta_3 ST + \beta_4 INI + \beta_5 SP + \beta_6 SD + \beta_7 SI + \beta_8 SC + \\ + \beta_9 DFI + \beta_{10} DGR + \beta_{11} DEQ + \beta_{12} DEE + \beta_{13} DSR + \beta_{14} DHU + \beta_{15} DSI + u$$

where the dependent variable π stands for FIE technological development (measured by our average FIE development index), *OP* for our index of operational autonomy, *MA* for our index of marketing autonomy, *ST* for our index of strategic autonomy, *INI* for our index of initiative for change, *SP* for the share of sales directed to the parent firm, *SD* for sales directed to the domestic market,[13] *SI* as the number of employees of the subsidiary, *SC* as the scope for a degree of product specialisation (measured in relative terms as the number of lines of business per employment of the FIE). In the second line, we add some dummies that we expect to influence our panels: the dummy *DFI* controls for whether the subsidiary produces final or intermediate products, *DGR* whether the subsidiary is a greenfield investment, *DEQ* whether the subsidiary is majority-owned by the parent (i.e. more than 50 per cent equity share), *DEE, DSR, DHU* and *DSI* are country dummies for Estonia, the Slovak Republic, Hungary, and Slovenia. Finally, we added a stochastic error term, *u*.[14]

The results of the stepwise OLS regression analysis turn out to be sufficiently robust to warrant reporting and interpretation (see Table 7.7). In the first model specification, we tested for all variables listed in our empirical regression formula. Here, we found three significant coefficients – for operational autonomy (*OP*), for the firm size (*SI*) and for the country dummy of the Slovak Republic (*DSR*). In the last and final model specification, the regression produces five significant coefficients – the one for operational autonomy (*OP*), for the firm size (*SI*), for the country dummy of the Slovak Republic (*DSR*), and additionally for the mandate indicator of sales to the parent company (*SP*) and for the scope of product specialisation (*SC*).

In our group of autonomy indices, the sign for operational autonomy (*OP*) is negative, which does not immediately correspond to our correlation analysis. However, we did establish that the correlation was extremely weak and statistically insignificant; moreover, in the regression model, a smaller number of firms took part in the analysis (N = 126 in model 1 and N = 186 in model 5 see Table 7.7) than in the correlation analysis (N = 442). More importantly, in view of the functional autonomy learning-curve of FIEs, this result is in fact plausible: operational autonomy is the first to increase with the maturing of subsidiaries, followed by the other two. The coefficients of the latter have the expected positive signs, albeit weaker and statistically insignificant. Throughout the five models, this result holds, with the coefficient slightly decreasing and the level of significance slightly increasing. The elasticity of increases in operational autonomy is hence 27 per cent for increases in the speed of FIE development. The initiative-related indicator bears the expected negative sign, yet remains insignificant with an error probability of some 86 per cent, which even increases in subsequent regression steps.

With respect to the mandate of subsidiaries, the coefficient for sales to parent companies turns significant only from model 3 – i.e. only after the exclusion of the distorting effects of other insignificant determinants. Still, an error probability of 8.5 per cent remains in the final model. The positive sign of the coefficient corresponds to the expecta-

Table 7.7 Results of a stepwise OLS regression in five model specifications (dependent variable: π FIE development)

Group	Variable	Model 1		Model 2		Model 3		Model 4		Model 5	
		Coefficient	*t*-value	Coefficient	*t*-value	Coefficient	*t*-value	Coefficient	t-value	Coefficient	*t*-value
	Constant	*** –1.39	–3.35	*** –1.38	–3.49	*** –1.45	–4.53	*** –1.68	–5.97	*** –1.68	–5.97
Functional autonomy	*OP*	*** –0.27	–2.57	*** –0.26	–3.31	*** –0.25	–3.25	*** –0.23	–3.23	*** –0.21	–3.07
	MA	0.02	0.25								
	ST	0.01	0.07								
Initiative for change	*INI*	–0.03	–0.18								
Mandate	*SP*	0.05	1.20	0.05	1.32	** 0.07	1.99	* 0.06	1.76	* 0.06	1.74
	SD	–0.02	–0.46	–0.03	–0.64						
Firm-specific variables	*SI*	*** 0.11	2.30	*** 0.11	2.42	*** 0.10	2.40	*** 0.11	2.77	*** 0.12	2.94
	SC	0.05	1.32	0.05	1.48	0.05	1.47	0.05	1.46	** 0.06	2.00
Firm-specific dummies	*DFI*	–0.00	–0.03								
	DGR	0.08	0.59	0.08	0.63						
	DEQ	–0.15	–0.73	–0.16	–0.85	–0.16	–0.88				
Country dummies	*DEE*	0.13	0.67	0.13	0.70	0.17	1.00	0.21	1.32		
	DSR	* –0.40	–1.87	* –0.40	–1.91	* –0.27	–1.90	–0.21	–1.55	* –0.23	–1.75
	DHU	–0.09	–0.51	–0.09	–0.53						
	DSI	–0.22	–1.28	–0.22	–1.33	–0.13	–1.08				
Regression	*corr.* R^2	0.07		0.10		0.12		0.12		0.12	
	N	126		144		186		186		186	

Note: Aster ISNS denote levels of statistical significance: *** at the 1% level of error probability, ** at the 5% level, * at the 10% level.
Source: CEE subsidiary database.

tions from our correlation analysis, where average FIE development was slightly higher for FIEs with a mandate that is closely tied to servicing the parent firm (i.e. extended workbench or assembly point), than in the case of a more independent mandate. The magnitude of elasticity of increases in mandate tightness for increases in the speed of FIE development is also very low indeed, at about 6 per cent. Our mirror indicator for sales to the domestic market does bear the expected negative sign, yet remains insignificant. The error probability falls slightly in subsequent steps, but insufficiently, and so is excluded from model 3 onwards.

Among the newly included firm-specific variables and firm-specific dummies, only the coefficients for size and product scope turn out significant: increases in subsidiary size display a positive elasticity for increases in the speed of FIE development throughout all the model specifications, which amounts to 12 per cent in the final specification. The respective elasticity for increases in the scope of products produced by the subsidiary reaches about half this magnitude, and becomes statistically significant only in the final model specification. The firm-specific dummies, controlling for final product producers, greenfield investments, and majority equity share subsidiaries, all turn out not to significantly influence FIE development in our panel. *T*-values for the latter do increase in the subsequent regression steps, but those increases remain insufficient for our statistical analysis.

Most of the country dummies remain statistically insignificant, only that for the Slovak Republic indicating that the country's foreign investment subsidiaries behave differently. This result suggests the need for further exploration in cross-country analysis in future research.

R^2 are generally very low which, however does not diminish the quality of results: we are not interested in generating a model able to exhaustively explain the sources of technology transfer (and so including a critical amount of explanatory variables). Rather, we want to infer from this regression analysis what elasticity our determinants reach respectively in the selection of variables we have accounted for in our analysis.

This suggests that FIE development is influenced most strongly by increases in the operational autonomy of subsidiaries, followed by the size of the subsidiary. The subsidiary's mandate *vis-à-vis* its parent company and the product scope also determine FIE development, yet to a much lower extent. We can thus conclude, that – at least for the members in our panel – FDI subsidiaries in CEE can increase their potential to benefit from technology transfer (a) by gaining (or being granted) more autonomy in operational business functions (such as decisions taken on accounting and finance, supply and logistics, operational management and process engineering); (b) by focusing their production more closely on their parent companies; (c) by increasing their scope of different lines of business; and (d) finally, by increasing their employment size.

Notes

1. The conceptual framework developed here is derived from the methodological approach developed for the field work by Slavo Radošević (for a more comprehensive explanation of the framework, see Majcen, Radošević and Rojec, 2003c: 10–13).
2. See, e.g. White and Poynter (1984); Young, Hood and Dunlop (1988); Bartlet and Ghoshal (1989); Birkinshaw and Hood (1998); Tavares (2001) and Holm, Malmberg and Sölvell (2002).
3. This corresponds to the short-term and long-term impacts as conceptualised by von Tunzelmann (2004) in his discussion of 'network-alignment'. Compare this interpretation with, e.g., Moran and Bergsten, 1998 and, for 'open networking' or 'strategic technology transfer', Dyker and von Tunzelmann, 2001.
4. In Szalavetz (2000), this link between a change in the level of autonomy and the slope of the learning curve is conceptualised by distinguishing between the static and dynamic modernisation effects of FDI. Here, static modernisation effects relate to low autonomy in all but operational functions and lead the FIE to achieve production capability and similar efficiency levels as in the parent company. Unless the autonomy position of the FIE is upgraded, FIE growth (in sales, exports, etc.) will remain static. The dynamic effects of FIE development set in only when the subsidiary assumes responsibility for additional business functions (functional upgrading).
5. With the subsidiary forming the subject of our field study, adaptive capacities were examined only at the subsidiary level. The host economy's absorptive capacity, a further determinant of technology transfer, does not form part of our analysis.
6. In the Hungarian case, the most important FIEs were not willing to participate in the survey. This reluctance resulted in our selection of Hungarian FIEs being less export oriented and weaker in capital and sales than the whole population of manufacturing FIEs. The distribution of sample FIEs by industries, however, is fully representative.
7. From our questionnaire, we translated FIE managers' answers according to the following formula: decisions taken by the FIE only (0), taken mainly by the FIE (0.33), taken mainly by the parent company (0.67) and taken by the parent company only (1). With respect to productivity growth as our main indicator for FIE development are used the formula: considerable reduction (–1), reduction (–0.5), no change (0), increase (0.5) and considerable increase (1).

8. The results of our rank correlation not only inform us about the relevance of the indicator of FIE technological position (the size of the correlation coefficient) and the direction of the relationship (the sign of the correlation coefficient), but also whether the correlation coefficient is in fact significant across the groups of FIEs.
9. Even though this classification scheme is widely used in the empirical literature, there are important difficulties involved. This classification is often at a two-digit NACE level which is not always homogeneous with respect to the technology intensity of all the member industries. In addition, we can assume that foreign investors in sectors which are typically considered to exhibit a high technology intensity might tend to allocate the less technology-intensive parts of their production chain to host countries with lower wages.
10. However, the *acquis communautaire* demands the strict adherence to high standards in the food industry. This might necessitate a closer control of subsidiaries in this industry, despite its low technological intensity.
11. Arguably, in some branches of Hungarian manufacturing, the term 'dual economy' could be a misleading one, when the branch is somewhat over-dominated by foreign investments, and little national activity remains.
12. Due to the research methodology of field work, our indicators are not measured in continuous (directly measurable) terms but rather in terms of ranks according to the perception of FIE managers. In our correlation analysis, we treat the individual answers as ranks, and hence use bivariate Spearman-Rho correlations.
13. There is no heteroscedasticity problem between sales directed to the parent company and sales directed to the domestic market, because a large share of sales are directed towards 'other external markets', not included in either of the two categories above. This share amounted on average to about 21 per cent across all our FIEs.
14. The country dummies have to include all countries except one; in our case, we selected Poland as the biggest. This is because we are interested in a general model. Hence, we use a single constant and country dummies. A model with fixed effects for each country or for FIE groups belonging to certain sectors could be expected to produce interesting additional results. However, this is beyond the scope of this chapter and will be assessed in future analysis of the database.

Appendix The Questionnaire

Number: _ _ _

QUESTIONNAIRE FOR FOREIGN INVESTMENT ENTERPRISES

1. What is **your firm's activity** at 3-digit NACE code (See attached classification. If your registration is very broad, please indicate the most important activities only):
   _ ________________; _ ________________; _ ________________

2. What is the total **number of employees** employed in your company (in 2001): _ _ _ _ _

3. What is the **year of establishment** of your company: _ _ _ _

4. What is the year of registration of your company **as a foreign investment enterprise** or announcement of foreign owner to the Firm Registration Office (if different from question 3): _ _ _ _

5. What is the current **equity share of the foreign owner**?
   ☐ 10–50% ☐ 51–99% ☐ 100%

6. Did you produce ☐ **intermediary** goods, ☐ **final** products, or ☐ both (in 2001)?[a]

7. Please, tick ☑ which **business functions are being undertaken** (a) on your own only, (b) mainly on your own, (c) mainly by your foreign owner, or (d) by your foreign owner only.

Business functions	only your company	mainly your company	mainly foreign owner	only foreign owner
Product development[b]	☐	☐	☐	☐
Process engineering[c]	☐	☐	☐	☐
Determining the product price	☐	☐	☐	☐
Supply and logistics	☐	☐	☐	☐
Accounting and finance of operations	☐	☐	☐	☐
Investment finance	☐	☐	☐	☐
Market research[d]	☐	☐	☐	☐
Distribution, sales	☐	☐	☐	☐
After-sale services	☐	☐	☐	☐
Advertisement[e]	☐	☐	☐	☐
Marketing[f]	☐	☐	☐	☐

Business functions	only your company	mainly your company	mainly foreign owner	only foreign owner
Operational management[g]	☐	☐	☐	☐
Strategic management or planning	☐	☐	☐	☐

8. How many **lines of businesses**[h], or clearly different product lines, did you produce at the time of establishing your foreign investment enterprise (as in question 4)? ___

   How many lines of businesses do you currently have? ___

9. Please evaluate the **magnitude of the changes** of categories below since the registration of your company as a foreign investment enterprise (as in question 4). Please tick ✓ appropriately:

– 2 = considerable reduction; – 1 = reduction; 0 = no change; 1 = increase; 2 = considerable increase

Value of total sales	–2	–1	0	1	2
Share of exports	–2	–1	0	1	2
Level of productivity in production	–2	–1	0	1	2
Level of technology of production equipment	–2	–1	0	1	2
Level of quality of produce	–2	–1	0	1	2

10. Please indicate the **structure and value of your sales (in %)** according to the origin of buyers:

Sales to your foreign owner	
Sales to other foreign buyers	
Sales to other domestic subsidiaries of your foreign owner	
Sales to other domestic buyers	

11. Please indicate the **structure of the value of your supplies (in %)** according to the items below:

Imports from your foreign owner	
Imports from other foreign suppliers	
Supplies from other domestic subsidiaries of your foreign owner	
Supplies from other domestic suppliers	
TOTAL	100%

12a. How important are **each of the following areas** for your competitiveness? Please indicate the appropriate number in each cell of the table.

1 = not important; 2 = not very important; 3 = important; 4 = very important; 5 = extremely important

Quality control assistance	Patents and licenses, R&D	People and training	Management

12b. How important are **the following sources for individual areas** of competitiveness? Please indicate the appropriate number in each cell of the table.[i]

1 = not important; 2 = not very important; 3 = important; 4 = very important; 5 = extremely important

SOURCES \ AREAS	Quality control assistance	Patents licenses, R&D	People and training	Management and
Your own organisation				
Your foreign owner company				
Other buyers abroad				
Other sellers abroad				
Other domestic subsidiaries of your foreign owner company				
Other buyers at home				
Other sellers at home				
Other sources (R&D institutes, universities, consultancies, etc.)				

13. Please indicate ✓ how important is **each of the following sources of finance** for your company?

1 = not important; 2 = not very important; 3 = important; 4 = very important; 5 = extremely important

Your retained earnings	1	2	3	4	5
Your foreign owner company	1	2	3	4	5
Other foreign sources (banks, other firms, etc.)	1	2	3	4	5
Other domestic subsidiaries of your foreign owner	1	2	3	4	5
Other domestic sources (banks, other firms, etc.)	1	2	3	4	5

14. Who has undertaken the **initiative for changes in the following areas:**

	only your company	mainly your company	mainly foreign owner	only foreign owner
In organization of business functions[i]				
In number of lines of businesses				
In sales and exports				

15. Please, tick ☑ directions in which you expect that **your mandate will evolve?**

	Increase	Decrease	Unchanged
Sales and exports			
Number of other business functions undertaken independently			
Number of lines of business (line of products)			

Background information and definitions

Notes

1 A 'Foreign investment enterprise' is a company with minimum of 10% of shares in foreign ownership.
2 A 'Foreign owner' or 'Foreign owner company' is the owner of the foreign equity share.
3 Please fill in the questionnaire with data related to the year 2001.

Some additional explanations for the questions:

a Final products are those produced directly for the final consumer market; intermediary goods constitute goods or materials which are used as in input for further production and are typically sold to other producing firms.
b Technical product development entails the development of the product in terms of what functions the product provides as well as the technical solutions to be solved to allow the product to offer those functions.
c Process engineering includes activities geared towards finding an efficient way to organise the process of production.
d Market research for the product is an assessment of the expectable quantitative extent of demand for the product and an assessment of customer preferences for the design of the product. The design incorporates both the range of possibilities for which the product can be used and the way the product looks like (colour, shape, etc.).
e The main focus of organisation of advertisement is on the development of means of advertisement (in media, on the street, in public transport, advertisement campaigns, etc.) and not on the financing of such activities.

f Marketing entails not only advertisement activities (as above) but also all activities within the company which aim at increasing the demand for the product (e.g. search for markets, changes to the product according to the preferences of the customers, etc.).

g Operational management (or short-term planning) includes activities geared towards the day-to-day operations of the company.

h The number of lines of business is the number of clearly different products you produce. Products belong to a different business line if they offer significantly different services to the customer, or are targeted at clearly different consumers. For example, production processes for wines of different qualities, which serve different market segments, represent different lines of business.

i The table is designed in a matrix form. That means that you should take into account both rows and columns when determining your answer in *each* cell. In question 12, we want to know e.g. in the first cell of the table how important for your company is the source of 'Your own organisation' for the area of competitiveness of 'Quality control assistance'. Please enter one value between 1 = not important and 5 = extremely important in that cell. Do the same in *all other* cells.

j Business functions include procurement, sales, marketing, production, R&D, engineering, maintenance, after-sale services, finance, accounting, strategic planning, etc.

Bibliography

Aitken, B. and A.E. Harrison (1999) 'Do Domestic Firms Benefit from Direct Foreign Investment? Evidence from Venezuela', *American Economic Review*, 89(3), pp. 605–18.

Aitken, B., A.E. Harrison and R.E. Lipsey (1996) 'Wages and Foreign Ownership: A Comparative Study of Mexico, Venezuela, and the United States?', *Journal of International Economics*, 40(3–4), pp. 345–71.

Altenburg, T. (2000) 'Linkages and Spillovers between Transnational Corporations and Small and Medium-Sized Enterprises in Developing Countries: Opportunities and Policies', Working Paper, 5/2000, Berlin: German Development Institute.

Alverez, I., J.P. Damijan and M. Knell (2002) 'Do Spanish Firms Get Technology through FDI and Trade?', University of Madrid, mimeo.

Anderson, T. and C. Hsiao (1982) 'Formulation and Estimation of Dynamic Models using Panel Data', *Journal of Econometrics*, 18, pp. 47–82.

Arellano, M. and S.R. Bond (1991) 'Some Tests of Specifications of Panel Data: Monte Carlo Evidence and an Application to Employment Equations', *Review of Economic Studies*, 58, pp. 277–94.

Arellano M. and S.R. Bond (1998) 'Dynamic Panel Data Estimation using DPD98 for GAUSS', London: Institute for Fiscal Studies, mimeo.

Arnold, J. (2003) 'Productivity Estimation at the Plant Level', 1st draft, Bocconi University, mimeo.

Bank of Estonia (2004) *Balance of Payments of Estonia*, http://www.eestipank.info.

Baltagi, H.B. (2001) *Econometric Analysis of Panel Data*, 2nd edn (Chichester: John Wiley).

Bartlett, C.A. and S. Ghoshal (1989) *Managing Across Borders: The Transnational Solution* (Boston: HBS Press).

Barrel, R. and D. Holland (2000) 'Foreign Direct Investment and Enterprise Restructuring in Central Europe', *Economics of Transition*, 8(2), pp. 477–504.

Barrel, R. and N. Pain (1997) 'Foreign Direct Investment, Technological Change, and Economic Growth within Europe', *Economic Journal*, 107, pp. 1770–86.

Barry, F., Gorg, H. and E. Strobl (2001) 'Foreign Direct Investment and Wages in Domestic Firms: Productivity Spillovers vs. Labour Market Crowding Out', Mimeo, University College Dublin and University of Nottingham.

Bernard, A.B. and J.B. Jensen (1999) 'Exceptional Exporter Performance: Cause, Effect or Both?', *Journal of International Economics*, 47, pp. 1–25.

Bessonova, E., K. Kozlov and K. Yudaeva (2003) *Trade Liberalization, Foreign Direct Investment, and Productivity of Russian Firms*, CEFIR Conference on 'Negotiating Russia's WTO Accession: Strategic Lessons from Multilateral Trade Liberalization and Club Enlargement', September.

Birkinshaw, J. (1998) 'Corporate Entrepreneurship in Network Organizations: How Subsidiary Initiative Drives Internal Market Efficiency', *European Management Journal*, 16(3), pp. 355–64.

Birkinshaw, J. and Hagtsröm (2000) *The Flexible Firm – Capabilities in Network Organisations* (New York: Oxford University Press).

Birkinshaw, J. and N. Hood (eds) (1998) *Multinational Corporate Evolution and Subsidiary Development* (New York: St Martin's Press).

Birkinshaw, J. and N. Hood (1999) 'Multinational Subsidiary Evolution: Capability and Charter Change in Foreign-Owned Subsidiary Companies', *Academy of Management Review*, 23(4), pp. 773–95.

Blalock, G. (2001) 'Technology from Foreign Direct Investment: Strategic Transfer Through Supply Chains', Haas School of Business, University of California, Berkeley, mimeo.

Blalock, G. and J. Gertler (2004) 'Welfare Gains from FDI through Technology Transfer to Local Suppliers', University of California, Berkeley, mimeo.

Blomström, M. and A. Kokko (1996) 'The Impact of Foreign Investment on Host Countries: A Review of the Empirical Evidence', Stockholm: Stockholm School of Economics.

Blomström, M. and A. Kokko (1997) 'How Foreign Investment Affects Host Countries', Policy Research Working Paper, 1745, World Bank.

Blomström, M. and A. Kokko (2002) 'FDI and Human Capital: A Research Agenda', Technical Paper, 195, OECD Development Centre, Paris: OECD.

Blomström, M. and A. Kokko (2003a) 'Human Capital and Inward FDI', SSE Working Paper Series, Working Paper, 167.

Blomström, M. and A. Kokko (2003b) 'The Economics of Foreign Direct Investment Incentives', Stockholm School of Economics Working Paper, 168.

Blomström, M. F. and F. Sjöholm, (1999) 'Technology Transfer and Spillovers: Does Local Participation with Multinationals Matter?', *European Economic Review*, 43, pp. 915–23.

Blomström, M. and E. Wolff (1994) 'Multinational Corporations and Productivity Convergence in Mexico', in W. Baumol, R. Nelson and E. Wolf (eds), *Convergence of Productivity: Cross-National Studies and Historical Evidence* (Oxford: Oxford University Press), pp.115–43.

Blomström, M., A. Kokko and M. Zejan (1994) 'Host Country Competition, Labour Skills and Technology Transfer by Multinationals', *Weltwirtschaftliches Archiv*, 128, pp. 522–33.

Blomström, M. *et al.* (1994) *Transnational Technology* (New York: United Nations).

Blundell, R.W. and S.R. Bond (1998) 'Initial Conditions and Moment Restrictions in Dynamic Panel Data Models', *Journal of Econometrics*, 87, pp. 115–43.

Blundell, R.W. and S.R. Bond (1999) 'GMM Estimation with Persistent Panel Data: An Application to Production Functions', Working Paper, W99/4, London: Institute for Fiscal Studies.

Borensztein, E., J. De Gregorio, and J.-W. Lee (1998) 'How Does Foreign Direct Investment Affect Economic Growth?', *Journal of International Economics*, 45, pp. 115–35.

Bosco, M.G. (2001) 'Does FDI Contribute to Technological Spillovers and Growth? A Panel Data Analyses of Hungarian Firms', *Transnational Corporations*, 10, pp. 43–68.

Breschi, S. and F. Lissoni (2001) 'Knowledge Spillovers and Local Innovation Systems: A Critical Survey', Liuc Papers, 84, Serie Economia e Impresa.

Buckley, P.J., J. Clegg and C. Wang (2002) 'The Impact of Inward FDI on the Performance of Chinese Manufacturing Firms?', *Journal of International Business Studies*, 33(4), pp. 637–55.

Campos, N. and Y. Kinoshita (2002) 'Foreign Direct Investment as Technology Transferred: Some Panel Evidence from the Transition Countries', WDI Working Paper, 438, Michigan.

Cantwell, J. and G.D. Santangelo (1999) 'The Frontier of International Technology Networks: Sourcing Abroad the Most Highly Tacit Capabilities', *Information Economics and Policy*, 11, pp. 101–23.

Caves, R.E. (1974) 'Multinational Firms, Competition and Productivity in Host-Country Markets?', *Economica*, 41(162), pp. 176–93.

Caves, R.E. (1996) *Multinational Enterprise and Economic Analysis* (Cambridge: Cambridge University Press).

CEE subsidiary database, http://www.iwh-halle.de/projects/productivity-gap/.

Child, J. (1993) 'Society and Enterprise between Hierarchy and Market', in J. Child (ed.), *Societal Change between Market and Organization* (Aldershot: Avebury).

Chudnovsky, D., A. López and G. Rossi (2003) 'Foreign Direct Investment Spillovers and the Absorptive Capabilities of Domestic Firms in the Argentine Manufacturing Sector in the 90s', Paper prepared for the 1st GLOBELICS Conference, Rio de Janeiro.

Clerides, S., S. Lach and J. Tybout (1997) 'Is Learning-by-Exporting Important? Micro-Dynamic Evidence from Colombia, Mexico and Morocco', *Quarterly Journal of Economics*, 113, pp. 903–47.

Coase, R.H. (1937) 'The Nature of the Firm', *Economica*, 4, pp. 386–405.

Cohen, W. and D. Levinthal, (1989) 'Innovation and Learning: The Two Faces of R&D', *Economic Journal*, pp. 569–96.

Country studies (2003) 'Mechanisms of Productivity Growth through Inward Foreign Direct Investment', http://www.iwh-halle.de/projects/productivity-gap/:

Estonia: Hannula, H., K. Männik and U. Varblane, Faculty of Economics and Business Administration, University of Tartu, mimeo.

Hungary: Hamar, J., A. Béres and Á. Mészáros, Kopint-Datorg Foundation for Economic Research, mimeo.

Poland: Kosmider, R., K. Malik and M. Kania, Politechnika Opolska, Faculty of Management and Production Engineering, mimeo.

Slovenia: Majcen, B. and M. Rojec, Institute of Economic Research, Ljubjana, mimeo.

Slovakia: Sabol, T., A. Hoškova, M. Šoltés and J. Hreňo, Technical University of Kosice, Faculty of Economics, mimeo.

Damijan, J.P. and M. Knell (2003) 'Impact of Privatization Methods on the Accessibility of Local Firms to International Knowledge Spillovers through Trade and Foreign Investment: Evidence from Estonia and Slovenia', University of Oslo, mimeo.

Damijan, J.P., M. Knell, B. Majcen and M. Rojec (2003a) 'The Role of FDI, R&D Accumulation and Trade in Transferring Technology to Transition Countries: Evidence from Firm Panel Data for Eight Transition Countries', *Economic Systems*, 27, pp. 189–204.

Damijan, J.P., M. Knell, B. Majcen and M. Rojec (2003b) 'Technology Transfer through FDI in Top-10 Transition Countries: How Important are Direct Effects, Horizontal and Vertical Spillovers?', WDI Working Paper, 549.

Damijan, J. P., B. Majcen, M. Knell and M. Rojec (2003c) 'Technology Transfer through FDI in Top-10 Transition Countries: How Important are Direct Effects, Horizontal and Vertical Spillovers?', Working Paper, 17, IER, University of Ljubljana.

De Backer, K. and Sleuwaegen, L. (2003) 'Does Foreign Direct Investment Crowd out Domestic Entrepreneurship?', *Review of Industrial Organization*, 22, 1, pp. 67–84.

Delgado, M.A., J.C. Farinas and S. Ruano (2001) 'Firm Productivity and Export Markets: A Non-Parametric Approach', *Journal of International Economics*, 57, pp. 397–422.

Djankov, S. and B. Hoekman, (1998) 'Avenues of Technology Transfers: Foreign Investment and Productivity Change in the Czech Republic', CEPR Discussion Paper, 1883.

Djankov, S. and B. Hoekman (2000) 'Foreign Investment and Productivity Growth in Czech Enterprises', *World Bank Economic Review*, 14(1), pp. 49–64.

Djankov, S. and P. Murrell (2002) 'Enterprise Restructuring in Transition: A Quantitative Survey', *Journal of Economic Literature*, 9, pp. 739–91.

Driffield, N. (2001) 'The Impact on Domestic Productivity of Inward Investment in the UK', Working Paper, 69(1), The *Manchester School*, 69(1), pp. 103–19.

Driffield, N. and J. Love (2003) 'Foreign Direct Investment, Technology Sourcing and Reverse Spillovers', *The Manchester School*, 71, pp. 659–72.

Dunning, J.H. (1977) 'Trade, Location of Economic Activity and MNE: A Search for an Eclectic Approach', in B. Ohlin, P.-O. Hesselborn and P.M. Wijkman (eds), *The International Allocation of Economic Activity* (London: Macmillan), pp. 395–418.

Dunning, J.H. (1981) *International Production and the Multinational Enterprise* (London: George Allen & Unwin).

Dunning, J.H. (1988) 'The Eclectic Paradigm of International Production: A Restatement and Some Possible Extensions', *Journal of International Business Studies*, 19(1), pp. 1–31.

Dunning, J.H. and R. Narula (1995) 'The R and D Activities of Foreign Firms in the United States', *International Studies of Management and Organization*, 25(1–2), pp. 85–103.

Dyker, D.A. (ed.) (1999) *Foreign Direct Investment and Technology Transfer in the Former Soviet Union* (Cheltenham: Edward Elgar).

Dyker, D.A. and C. Stolberg (2003) 'Productivity and Capability in the Transition Economies: A Historical and Comparative Perspective', Brighton: University of Sussex, mimeo.

Dyker, D.A. and N. von Tunzelmann (2002) 'Network Alignment in Firms in Transition Countries: A Survey of Hungarian and Slovenian Companies', University of Sussex, mimeo.

Eaton, J. and S. Kortum (1996) 'Trade in Ideas: Patenting and Productivity in the OECD', *Journal of International Economics*, 40, pp. 251–71.

EBRD (2004) Transition Report Update, April, European Bank for Reconstruction and Development, London.

Egelhoff, W., L. Gorman and S. McCormick (1998) 'Using Technology as a Path to Subsidiary Development', in J. Birkinshaw and N. Hood (eds), *Multinational Corporate Evolution and Subsidiary Development* (New York: St Martin's Press), pp. 213–38.

EUROSTAT (2003) Economy and Finance outline database (www.europa.en.int/comm/eurostat)

Evenett, S.J. and A. Voicu (2002) 'Picking Winners or Creating Them? Revisiting the Benefits of FDI in the Czech Republic?', World Bank, mimeo.

Feenstra, R., J. Markusen and W. Zeile (1992) 'Accounting for Growth with New Inputs: Theory and Evidence', *American Economic Revue*, 82, pp. 415–21.

Felipe, J. (1999) 'Total Factor Productivity Growth in East Asia: A Critical Survey', *Journal of Development Studies*, 35(4), pp. 1–41.

Findlay, R. (1978) 'Relative Backwardness, Direct Foreign Investment, and the Transfer of Technology: A Simple Dynamic Model', *Quarterly Journal of Economics*, 92, pp. 1–16.

Foreign Investor (2000) 'Foreign Direct Investments in Slovenia'.

Frydman, R., Gray, C., Hajel, M. and Audrzej Repaoyuski (1997) *Private Ownership and Corporate Performance: Evidence from Transition Economies*, EBRD Working Paper, no. 26.

Gershenkron, A. (1952) 'Economic Backwardness in Historical Perspective', in B.F. Hoselitz (ed.), *The Progress of Underdeveloped Areas* (Chicago: University of Chicago Press).

Gestrin, M. (2001) 'Is Export-Oriented FDI Better?', OECD Global Forum on International Investment: New Horizons and Policy Challenges for Foreign Direct Investments in the 21st Century, Mexico, pp. 1–9.

Girma, S. and K. Wakelin (2000) 'Are There Regional Spillovers from FDI in the UK?', GEP Research Paper, 00/16, Nottingham: University of Nottingham.

Glass, A.J. and K. Saggi (2002) 'Multinational Firms and Technology Transfer', *Scandinavian Journal of Economics,* 104(4), pp. 495–513.

Globerman, S., A. Kokko and F. Sjöholm (1996) 'Technology Sourcing in Swedish MNEs and SMEs: Evidence from Patent Data', Working Paper Series in Economics and Finance, 125, Stockholm: Stockholm School of Economics.

Görg H. and D. Greenaway (2001) 'Foreign Direct Investment and Intra-Industry Spillovers: A Review of the Literature', GEP Research Paper, 01/37, Nottingham: University of Nottingham.

Görg, H. and E. Strobl (2001) 'Multinational Companies and Productivity Spillovers: A Meta-Analysis with a Test for Publication Bias', *Economic Journal,* 111(475), pp. 723–39.

Greene, W. H. (1993) *Econometric Analysis*, 2nd edn (Englewood Cliffs, NJ: Prentice-Hall International)

Greenaway, D., N. Sousa and K. Wakelin (2004) 'Do Domestic Firms Learn to Export from Multinationals?', *European Journal of Political Economy*, 20, pp. 1027–44..

Griffith, R. (1999) 'Using the ARD Establishment-Level Data to Look at Foreign Ownership and Productivity in the UK', Institute for Fiscal Studies, Working Paper, 99/06, London: Institute for Fiscal Studies.

Griffith, R. (2004) 'Foreign Ownership and Productivity: New Evidence from the Service Sector and the R&D Lab', Institute for Fiscal Studies, Working Paper, 04/22, London: Institute for Fiscal Studies.

Griliches, Z. and J. Mairesse (1995) 'Production Functions: The Search for Identification', NBER Working Paper, 5067, Cambridge, MA: National Bureau of Economie Research.

Grossman, G. and E. Helpman (1991) *Innovation and Growth in the Global Economy* (Cambridge, MA: MIT Press).

Haddad, M. and A. Harrison (1993) 'Are There Positive Spillovers from Direct Foreign Investments? Evidence from Panel Data for Morocco', *Journal of Development Economics*, 42, pp. 51–74.

Hannula, H. and K. Tamm (2003) 'Restructuring and Efficiency in the Estonian Manufacturing Industry: The Role of Foreign Ownership', Working Paper Series of Tartu University, 15, Faculty of Economics and Business Administration.

Hardin, J. (1997) 'Obtaining Mill's Ratio for Heckman Models', Stata FAQs, College Station, Texas: StataCorp LP.

Harris, R. and C. Robinson (2001) 'Foreign Ownership and Productivity in the United Kingdom: Estimates for UK Manufacturing Using the ARD', *Review of Industrial Organization*, 22(3), pp. 207–23.

Haskel, J.E., S.C. Pereira and M.J. Slaughter (2001) 'Does Inward Foreign Direct Investment Boost the Productivity of Domestic Firms?', NBER Working Paper, 8433, Cambridge, MA: National Bureau of Economic Research.

Hatzichronoglou, T. (1997) *Revision of the High-technology Sectors and Product Classification*, OECD STI Working Paper, 2 (Paris: OECD).

Havrylyshyn, O. (2001) 'Recovery and Growth in Transition Countries: A Decade of Evidence', IMF Staff Papers, 48 (Washington, DC: IMF).

Heckman, J.J. (1979) 'Sample Selection Bias as a Specification Error', *Econometrica*, 47, pp. 153–61.

Holm, U., A. Malmberg and Ö. Sölvell (2002) 'MNC Impact on Local Clusters', IIB Institute of International Business, Working Paper, 02/11, Stockholm: IIB.

Hunya, G. (1998) 'Integration of CEEC Manufacturing into European Corporate Structures via Direct Investment', WIIW Research Report, 245, Vienna: Vienna Institute for International Economic Studies.

Hunya, G. (ed.) (2000) *Integration through Foreign Direct Investment* (Cheltenham: Edward Elgar).

Hymer, S.H. (1976) *International Operations of National Firms: A Study of Direct Foreign Investment* (Cambridge, MA: MIT Press).

IMF (1993) *Balance of Payments Manual*, 5th edn (Washington, DC: International Monetary Fund).

IMF (2001) Managing Foreign Direct Investment (Washington, DC: International Monetary Fund).

Index of Economic Freedom (2003) The Heritage Foundation, *The Wall Street Journal*, p. 422.

Jovanovic, B. (1982) 'Selection and the Evolution of Industry', *Econometrica*, 50(3), pp. 649–70.

Jensen, C. (2002) 'Spillovers in the Polish Food Industry: Exploring the Role of Local Externalities and Global Networks', Paper presented at CEEC Workshop on Transition and Enterprise Restructuring, Copenhagen: CBS, 15–17 August, mimeo.

Judson, R.A. and A.L. Owen (1999) 'Estimating Dynamic Panel Data Models: A Guide for Macroeconomists', *Economics Letters*, 65, pp. 9–15.

Keller, W. (1996) 'Absorptive Capacity: On the Creation and Acquisition of Technology in Development', *Journal of Development Economics*, 49(1), pp. 199–227.

Kinoshita, Y. (2000) 'R&D and Technology Spillovers via FDI: Innovation and Absorptive Capacity', Working paper, 349, The William Davidson Institute, University of Michigan Business School.

Knickerbocker, F.T. (1973) *Oligopolistic Reaction and the Multinational Enterprise* (Cambridge, MA: Harvard University Press).

Kokko, A. (1992) 'Foreign Direct Investment, Host Country Characteristics, and Spillovers', Stockholm: The Economic Research Institute.

Kokko, A. (1994) 'Technology, Market Characteristics and Spillovers', *Journal of Development Economics*, 43, pp. 279–93.

Kokko, A., R. Tansini and M. Zejan (2001) 'Trade Regimes and Spillover Effects of FDI: Evidence from Uruguay', *Weltwirtschaftliches Archiv*, 137(1), pp. 124–9.

Kogut, B. and U. Zander (1993) 'Knowledge of the Firm and the Evolutionary Theory of the Multinational Corporation', *Journal of International Business Studies*, 24(3), pp. 625–46.

Konings, J. (2000) 'The Effects of Direct Foreign Investment on Domestic Firms: Evidence from Firm-Level Panel Data in Emerging Economies', Working Paper, 344, Michigan US: WDI.

Kosova, R. (2003) 'Do Foreign Firms Crowd Out Domestic Firms? The Evidence from the Czech Republic', University of Michigan Business School, mimeo.

Kugler, M. (2001) 'The Diffusion of Externalities from Foreign Direct Investment: Theory Ahead of Measurement', Discussion Papers in Economics and Econometrics, 23, Southampton: University of Southampton.

Krugman, P. (1991) *Geography and Trade* (Cambridge, MA: MIT Press).

Krugman, P. (1990) *The Age of Diminishing Expectations*, 1st edn (Cambridge, MA: MIT Press).

Krugman, P. and M. Obstfeld (2000) *International Economics: Theory and Policy* (Reading, MA: Addison-Wesley).

Kvinge, T. (2004) *Knowledge Diffusion through FDI – Established Wisdom or Wishful Thinking*? Working Paper, 37/2004. University of Oslo, Centre for Teohnology, Innovation and Culture.

Lane, P.J. and M. Lubatkin (1998) 'Relative Absorptive Capacity and Interorganizational Learning', *Strategic Management Journal*, 19(5), pp. 461–77.

Levin, R.C. and P.C. Reiss (1988) 'Cost-Reducing and Demand-Creating R&D with Spillovers', *Rand Journal of Economics*, 19(4), pp. 538–5.

Lim, E.-G. (2001) 'Determinants of, and the Relation between, Foreign Direct Investment and Growth: A Summary of the Recent Literature', IMF Working Paper, WP/01/175, Washington, DC: International Monetary Fund.

Liu, X. and Y. Wei (2003) 'Mutual Productivity Spillovers between Foreign and Local Firms in China', Aston Business School Research Paper, 0307, Aston: University of Birmingham.

Lutz, S.H. and O. Talavera (2003) 'Do Ukrainian Firms Benefit from FDI?', Discussion Paper, 03–05, Mannheim, Germany: Centre for European Economic Research.

Macjen, B., S. Radošević and M. Rojec (2003a) 'Strategic Control in Foreign Subsidiaries in Central European Countries', paper presented at the XI International Conference on European Studies 30 September–3 October, Havana.

Macjen, B., S. Radošević and M.Rojec (2003b) 'Productivity Growth and Functional Upgrading in FIEs in the Slovenian Manufacturing Sector', Working Paper, Ljubijan: Institute for Economic Research.

Majcen, B., S. Radošević and M. Rojec (2003c) 'FDI Subsidiaries and Industrial integration of Central Europe: Conceptual and Empirical Results', IWH Discussion Paper, 177, Halle: Halle Institute for Economic Research.

Mansfield, E. and A. Romeo (1980) 'Technology Transfer to Overseas Subsidiaries by US-Based Firms', *Quarterly Journal of Economics*, 95(4), pp. 737–50.

Markusen, J.R. (1984) 'Multinationals, Multi-Plant Economies, and the Gains from Trade', *Journal of International Economics*, 10(1984), pp. 205–26.

Markusen, J.R. (1989) 'Trade in Producer Services and in Other Specialized Intermediate Inputs', *American Economic Review*, 79, pp. 85–95.

Markusen, J.R. (1995) 'The Boundaries of Multinational Enterprises and the Theory of International Trade', *Journal of Economis Perspectives*, 9, pp. 169–89.

Markusen, J.R. and A.J. Venables (1997) 'Foreign Direct Investment as a Catalyst for Industrial Development', NBER Working Paper, 6241, Cambridge, MA: National Bureau of Economic Research.

McMillan, C.H. (1996) 'Foreign Direct Investment in Eastern Europe: Harnessing FDI to the Transition from Plan to Market', in S. Chan (ed.), *Foreign Direct Investment in a Changing Global Political Economy* (New York: St Martin's Press), pp. 127–49.

Meyer, K.E. (2000): 'International Production Networks and Enterprise Transformation in Central Europe', *Comparative Economic Studies*, 42(1), pp. 135–50.

Meyer, K.E. (2003) 'Foreign Direct Investment in Emerging Markets', DRC Working Paper, 15, Centre for New and Emerging Markets, London: London Business School, draft.

Meyer, K.E. (2004) 'Perspectives on Multinational Enterprises in Emerging Economies', *Journal of International Business Studies*, 35(4), pp. 259–76.

Moran, T.H. and C.F. Bergsten (1998) *Foreign Direct Investment and Development: The New Policy Agenda for Developing Countries and Economics of Transition* (Washington, DC: Institute for International Economics).

Mudambi, R. (1999) 'MNE Internal Capital Markets and Subsidiary Strategic Independence', *International Business Review*, 8(2), pp. 197–211.

OECD (1999) *OECD Science, Technology and Industry Scoreboard 1999: Benchmarking Knowledge-Based Economies* (Paris: OECD)

OECD (2002) 'Foreign Direct Investment in Slovenia: Trends and Prospects', *OECD Investment Policy Reviews: Slovenia*, Pans: OECD.

Okamoto, Y. (1999) 'Multinationals, Production Efficiency and Spillover Effects: The Case of the US Auto Parts Industry', *Weltwirtschaftliches Archiv*, 135(2), pp. 241–60.

Olley, S.G. and A. Pakes, (1996) 'The Dynamics of Productivity in the Telecommunications Equipment Industry', *Econometrica*, 64, pp. 1263–97.

Oulton, N. (1998) 'Labour Productivity and Foreign Ownership in the UK', NIESR Discussion Paper, 143, London: NIESR.

Ozawa, T. (1979) *Multinationalism, Japanese Style* (Princeton, NJ: Princeton University Press).

Paterson, S.L. and D.M. Brock (2001) 'The Development of Management Research: Review and Theoretical Analysis', *International Business Review*, 11, pp. 139–63.

Perez, T. (1997) 'Multinational Enterprises and Technological Spillovers: An Evolutionary Model', *Journal of Evolutionary Economics*, 7, pp. 169–92.

Perez, T. (1998) *Multinational Enterprises and Technology Spillovers*, (Amsterdam: Harwood Academic Publishers).

Rasiah, R. (1995) *Foreign Capital and Industrialisation in Malaysia* (New York: St Martin's Press).

Rodrik, D. (1999) 'The New Global Economy and Developing Countries: Making Openness Work', Policy Essay, 24, Washington, DC: Overseas Development Council/Johns Hopkins University Press.

Rojec, M. and M. Svetličič (2003) *Facilitating Transition by Internationalization: Outward Direct investment from Central European Economies in Transition* (Aldershot: Ashgate).

Roth, K. and J. Morrison (1992) 'Implementing Global Strategy: Characteristics of Global Subsidiary Mandates', *Journal of International Business Studies*, 23(4), pp. 715–35.

Sabol, T. (2003) Use of Knowledge Discovery Techniques in Evaluation of Foreign Direct Investment Survey', Technical University of Kosice, Faculty of Economics, Slovakia, mimeo.

Schoors, K. and B. van der Tool (2002) 'Foreign Direct Investment Spillovers within and between Sectors: Evidence from Hungarian Data', Working Paper, 157, Ghent: University of Ghent.

Sgard, J. (2001) 'Direct Foreign Investments and Productivity Growth in Hungarian Firms, 1992–1999, CEPII WP, Working Paper, 19, D: CEPII.

Sinani, E. and K.E. Meyer (2004) 'Identifying Spillovers from Technology Transfer from FDI: The Case of Estonia', *Journal of Comparative Economics*, 34(4), pp. 445–66.

Sjöholm, F. (1999) 'Technology Gap, Competition and Spillovers from Direct Foreign Investment: Evidence from Establishment Data', *Journal of Development Studies*, 36, pp. 53–73.

Slovenia (2004) *Your Business Partner 2003–2004*, Ljubljana: TIPO.

Smarzynska, B. (2002) 'Spillovers from Foreign Direct Investment through Backward Linkages: Does Technology Gap Matter?', Washington, DC: World Bank, mimeo.

Smarzynska-Javorcik, B. (2004) 'Does Foreign Direct Investment Increase the Productivity of Domestic Firms? In Search of Spillovers through Backward Linkages', *American Economic Review*, 94(3), pp. 605–27.

Smarzynska-Javorcik, B. and M. Spatareanu (2003) 'To Share or Not to Share: Does Local Participation Matter for Spillovers from FDI?', World Bank Policy Research Working Paper, 3118, Washington, DC: World Bank.

Sun, S. (2002) 'Is Firm Growth Proportional or Disproportional? A Reconciliation with an Application Using Franchising Data', Department of Economics, Ann Arbor, MI: University of Michigan, unpublished dissertation.

Szalavetz, A. (2000) 'Adjustment of Hungarian Engineering Companies to the Globalising Corporate Network', in Z. Bara and L. Csaba (eds), *Small Economies' Adjustment to Global Challenges* (Budapest: Aula Publishing), pp. 357–76.

Tavares, A.T. (2001) 'Strategic Management of Multinational Networks: A Subsidiary Evolution Perspective', Paper presented at the 6th International Symposium on International Manufacturing, University Cambridge, 9–11 September.

Torlak, E. (2004) 'Foreign Direct Investment, Technology Transfer, and Productivity Growth in Transition Countries Empirical Evidence from Panel Data', Göttingen: Centre for Globalisation and Europeanisation of the Economy, Georg-August-Universität.

Tunzelmann, N. von (2004) 'Network Alignment in the Catching Up Economies of Europe', in F. McGowan *et al.* (eds), *The Emerging Industrial Structure of the Wider Europe* (London: Routledge), pp. 23–37.

Vahter, P. (2004) 'The Effect of Foreign Direct Investments on Productivity in Estonian and Slovenian Manufacturing', University of Tartu, MA thesis.

Van den Bosch, F.A.J., H.W. Volberda and M. De Boer (1999) 'Co-evolution of Firm Absorptive Capacity and Knowledge Environment: Organizational Forms and Combinative Capabilities', *Organization Science*, 10(5), pp. 551–68.

Van Pottelsberghe, B. and F. Lichtenberg (2001) 'Does Foreign Direct Investment Transfer Technology across Borders?', *Review of Economics and Statistics*, 83(3).

Varblane, U. (2001) *Analysis of the Results of Survey Foreign Investor 2000* (Tartu: Enterprise Estonia).

Varblane, U. and T. Ziazic (2000) 'The Impact of Foreign Direct Investment on the Export Activities of Estonian Firms', *Journal of East West Business*, 5(1), pp. 173–90.

Varblane, U., T. Roolaht, E. Reiljan and R. Jüriado (2001) 'Estonian Outward Foreign Direct Investments', Working Paper Series of Tartu University, Faculty of Economics and Business Administration.

Veblen, T. (1915) *Imperial Germany and the Industrial Revolution* (London: Macmillan).

Vernon, R. (1966) 'International Investment and International Trade in the Product Cycle', *Quarterly Journal of Economics*, 2, p. 190–207.

Verspagen, B. (1999) 'A Global Perspective on Technology and Economic Performance, and the Implications for the Post-Socialist Countries', in D.A. Dyker and S. Radošević (eds), *Innovation and Structural Change in Post-Socialist Countries: A Quantitative Approach* (Dordrecht: Kluwer) pp. 29–44.

Wells, L.T. (1983) *Third World Multinationals* (Cambridge, MA: MIT Press).

White, R. and T. Poynter (1984) 'Strategies for Foreign Owned Subsidiaries in Canada', *Business Quarterly* 55(3), pp. 59–69.

Wolf, H.C. (1997) *Transition Strategies: Choices and Outcomes* (New York: Stern Business School), mimeo.

Wooldridge, J. (2002) *Econometric Analysis of Cross-Section and Panel Data* (Cambridge, MA: MIT Press).

World Investment Report (2000) *Cross-Border Mergers and Acquisitions and Development* (New York, Geneva: United Nations).

World Investment Report (2001) *Promoting Linkages* (New York, Geneva: United Nations).

World Investment Report (2002) *Transnational Corporations and Export Competitiveness* (New York, Geneva: United Nations).

World Investment Report (2003) *FDI Policies for Development: National and International Perspectives* (New York, Geneva: United Nations).

Young, S., N. Hood and S. Dunlop (1988) 'Global Strategies, Multinational Subsidiary Roles and Economic Impact in Scotland', *Regional Studies*, 22(6), pp. 487–97.

Zhang, K.H. (2001) 'Does Foreign Direct Investment Promote Economic Growth? Evidence from East Asia and Latin America', *Contemporary Economic Policy*, 19(2), pp. 175–85.

Zhigiang, L. (2002) 'Foreign Direct Investment and Technology Spillover: Euidence from China', *Journal of Comparative Economics*, 30, 579–602.

Zukowska-Gagelmann, K. (2001) 'Productivity Spillovers from Foreign Direct Investment in Poland', *Economic Systems*, 24(3), pp. 223–56.

Index